DATE DUE

AP 1 '96 AP 1 9 '96		
DE 2 '96		
DE 15 '03		
JE 7 '09		
DE 10 '19		

DEMCO 38-296

VGM Opportunities Series

OPPORTUNITIES IN
WASTE MANAGEMENT AND RECYCLING CAREERS

Mark Rowh

Foreword by
Dean L. Buntrock
Chairman and CEO
Waste Management, Inc.

VGM Career Horizons
a division of *NTC Publishing Group*
Lincolnwood, Illinois USA

Cover Photo Credits:
Front cover: upper left and lower left,
Waste Age; upper right, USDA Forest Service
photo by Yuen-Gi Yee; lower left, Reynolds
Metals Company.

Back cover: upper left and upper right,
Waste Age; lower left, Reynolds
Metals Company; lower right, Waste
Management Inc.

Library of Congress Cataloging-in-Publication Data

Rowh, Mark, 1952-
    Opportunities in waste management careers / Mark Rowh.

        p.   cm. — (VGM opportunities series)
    Includes bibliographical references.
    ISBN 0-8442-4018-4 (hardcover) ISBN 0-8442-4019-2
    (softcover)
        1. Refuse and refuse disposal—Vocational guidance. I. Title.
    II. Series
    TD793.R68 1992
    628.4′023—dc20   92-3243
                                                                                   CIP

Published by VGM Career Horizons, a division of NTC Publishing Group.
© 1992 by NTC Publishing Group, 4255 West Touhy Avenue,
Lincolnwood (Chicago), Illinois 60646-1975 U.S.A.

2 3 4 5 6 7 8 9 0 VP 9 8 7 6 5 4 3 2 1

ABOUT THE AUTHOR

Mark Rowh is a widely published writer as well as an experienced educator in community college and occupational education. He holds a doctorate in vocational and technical education from Clemson University, and currently serves on the administrative staff at New River Community College in Dublin, Virginia.

In his educational career, Rowh has worked closely with a variety of technical programs. He has taught and held administrative posts in both two-year and four-year colleges.

Rowh's articles on educational and management topics have appeared in more than fifty magazines. He is the author of eight books, including *Opportunities in Electronics Careers* and several other volumes in the VGM series.

ACKNOWLEDGMENTS

The author greatly appreciates the cooperation of the following in researching and writing this book:

American Water Works Association
Green River Community College
Institute of Scrap Recycling Industries
National Solid Wastes Management Association
Rutgers University
Solid Waste Association of North America
U.S. Department of Labor
U.S. Environmental Protection Agency
Waubonsee Community College

Some material has been adapted from *Occupational Outlook Quarterly* and the *Occupational Outlook Handbook,* published by the U.S. Department of Labor.

FOREWORD

Americans have become increasingly concerned in recent years about the quality of our environment and what we can do now to preserve the planet for future generations. Waste management professionals are at the forefront of this struggle to find sensible ways of disposing of waste. It is a large and diverse field that concerns itself with everything from toxic chemicals to ordinary household trash.

This book explores the roles of scientists, engineers, technicians, and businesspeople in the process of waste management. If the information you find here intrigues you, there are any number of careers you might eventually pursue. Perhaps you will become the director of your state recycling center, or a scientist conducting research for the Environmental Protection Agency. You might find work in the corporate sector helping businesses find safe and cost-efficient ways of disposing of industrial by-products.

Whatever your eventual niche in the industry, you will enjoy the satisfaction of knowing that your work helps provide the basic services essential in a civilized society, that it is key to preserving the quality of our environment now and in the future.

Dean L Buntrock
Chairman and CEO
Waste Management, Inc.

INTRODUCTION

Decisions about careers are among the most important choices anyone makes. An occupational choice made today will likely affect your options well into the twenty-first century. Accordingly, anyone considering career choices should base such decisions on information which is as complete as possible.

Each career field has its own particular advantages and disadvantages. In the area of waste management and recycling, the advantages can be significant, especially for people with the right interests, aptitudes and training. This is a career area offering a level of diversity which many might find surprising. At the same time, it represents a field with increasing potential—as our society becomes more and more concerned with protecting the environment and managing resources.

For anyone interested in considering a career in waste management or recycling, this book provides a basic overview of what jobs are performed and how one might prepare for them. The material should prove to be of help to anyone interested in learning more about this multifaceted career field.

DEDICATION

This book is dedicated to Dr. Henry Pate and the faculty of Clemson University.

CONTENTS

   The nature of waste management and recycling. More than meets the surface. Advantages
   of a career in waste management or recycling.
   Reviewing career possibilities in waste management.

   Taking the entry-level approach. Becoming a
   trained technician. Working as an engineer or

CHAPTER 1

IMPORTANCE OF WASTE MANAGEMENT AND RECYCLING

The next time you take out the garbage, stop a moment and think about what would happen if there were no structured system for removing waste materials. This might not be a tremendous problem if you live on a large ranch or farm, for it might be possible to bury waste products or otherwise dispose of them. But for residents of cities and towns, the lack of waste removal could pose serious problems. Yards, streets, and open areas might become filled with garbage, accompanied by plenty of bad smells, insects, and general ugliness. Without a doubt, it would not be a pretty sight!

The same is true of wastewater and biological waste. If these substances were not removed and treated appropriately, we would all be vulnerable not just to unpleasant surroundings but also to a host of dangerous diseases.

The need to deal with waste applies to the business and industrial sector as well as to private residences. In fact,

1

some of the most important work involved in waste management deals with the products and by-products of industry. For example, consider the following situations:

- A factory that produces chemical compounds used to make plastics accumulates tons of chemical wastes each hour it is in operation;
- A large hospital is faced with disposing of biological waste products such as tissues removed during surgery;
- A university which operates a small nuclear reactor must safely rid itself of radioactive waste materials.

Every day, large and small businesses across the United States and Canada, along with individuals and families, discard millions of pounds of waste material. Indeed, the overall scope of waste production is staggering.

In the United States alone, people discard the following each year, according to the Environmental Protection Agency (EPA):

- Over 30 million tons of yard waste
- More than 7.5 million tons of furniture
- Over 5 million tons of glass bottles
- About 4 million tons of junk mail
- Over 3 million tons of large appliances
- Over 5 million tons of books and magazines
- Nearly 3 million tons of disposable diapers
- About 3 million tons of paper tissue and towels
- Over 1.5 million tons of batteries
- Millions of tons of waste products of all kinds.

The accumulation of waste products means, among other things, that a substantial demand exists for workers who can dispose of them. Society needs people who can haul solid waste materials away, process wastewater, store refuse, recycle various materials, remove dangerous chemicals or biological wastes, and provide a range of related services. Thus the overall area of waste management provides a great variety of career potential.

THE NATURE OF WASTE MANAGEMENT AND RECYCLING

The term "waste management" refers to any organized process for removing, storing, or rendering harmless the waste products generated in a civilized society. "Recycling" takes matters a step further by taking discarded materials and processing them so that they can be used again.

The history of waste management is a long one, but it is only in recent years that recognition of the importance of managing waste effectively has become widespread.

Until modern times, there was little need for any formalized system for managing waste. Before the days of large-scale manufacturing and heavy concentrations of population, almost all materials were of biological origin and easily returned to nature. Valuable commodities such as metals were saved and even recycled, but such materials were the exception rather than the rule. As late as the eighteenth and nineteenth centuries, people regularly threw garbage in the streets, burned it, or simply ignored it.

With the industrial revolution, however, came the development of mass production. Then came more and more consumer products, huge amounts of industrial wastes, increasing human population, and the development of systems for managing the wastes produced by the various activities of society.

In the last two or three decades, a whole new emphasis on conserving resources and protecting the environment has emerged. People are more health-conscious than ever before, and the proper handling of waste is now seen as a matter of vital public interest. When this increased attention to health and environmental concerns is combined with the sheer increase in production of waste materials, the importance of waste management and recycling has grown to new heights.

MORE THAN MEETS THE SURFACE

The phrase "waste management" may not evoke the most positive of mental images when you first think about it. But think again. A career in this field can mean much more than the stereotypical and limited role of the person riding around on a truck and emptying garbage cans. While this is not to criticize people who hold such jobs, the field as a whole offers much more for those willing to undergo special training and to apply their skills and talents to specific job roles.

A career in waste management can mean working as a highly trained technician in a complex treatment plant. It can mean earning a degree in engineering and then specializing in working with hazardous waste materials. Or it can mean fulfilling a wide range of functions as a supervisor or manager. These and many other roles make up a field which offers a wide array of job opportunities.

The fact is, waste management and recycling represent extremely promising career areas for many individuals. The management of waste materials is vital to modern society, and its importance is expected to grow as populations increase, more and more waste materials are produced, and increasing attention is given to environmental concerns by governmental agencies, businesses and the public at large. Add to this the mushrooming interest in recycling, and it seems obvious that this overall area will be a focus of much attention—and a source of thousands of good jobs—through the remainder of the 1990s and beyond.

Subsequent chapters provide more details about specific job possibilities. The central point to remember here, though, is that waste management and recycling offer truly diverse occupational challenges. Working in this field might involve being employed by a municipal government, a private company, or even starting one's own small business. Jobs might be found in large cities or small towns. Job levels can range from those with minimal responsibilities to managing large numbers of workers or conducting complex technical tasks. With these and other options, careers in this field can meet a wide range of individual interests and occupational goals.

ADVANTAGES OF A CAREER IN WASTE MANAGEMENT OR RECYCLING

A number of factors contribute to the strong potential that waste management and recycling careers can offer. These include the following:

- Overall population growth means production of waste will continue to grow.
- Increased consumption of natural resources should lead to even more interest in recycling.
- Growing public concern about the environment should prompt government and industry to dedicate increasing amounts of attention—and resources—to waste management.
- New levels of understanding about human health will place added emphasis on proper handling and disposal of toxic substances.
- As the overall waste management industry thrives, many opportunities will be presented to men and women who decide to pursue careers in this area.

REVIEWING CAREER POSSIBILITIES IN
WASTE MANAGEMENT

The remainder of this book provides an overview of various facets of careers in recycling or waste management. Chapter 2 looks at the major types of jobs involved. Chapter 3 addresses the area of wastewater operations, and Chapter 4 covers solid waste management and recycling. Careers in hazardous waste management are discussed in Chapter 5.

The remaining chapters cover topics such as educational options, salaries and benefits, and advice on breaking into the field.

Several helpful appendixes are also included to provide additional information related to careers in this area, including a list of colleges offering programs of interest. If a career in waste management seems of interest, read on and see how you might fit in!

CHAPTER 2

DIVERSE CAREER POSSIBILITIES

The fields of waste management and recycling offer diverse career possibilities. Anyone who has not been exposed to this area may be surprised at the range of jobs involved. The level of diversity becomes less surprising, though, when you consider how more and more emphasis is being placed on protecting the environment, making the most of natural resources, and avoiding health problems which would occur if not for proper waste management practices.

Career areas range from jobs as laborers or other positions requiring little in the way of special training or skills, to jobs as highly trained scientists, engineers, or managers. Some of these career paths are described below, with more details covered in subsequent chapters.

TAKING THE ENTRY-LEVEL APPROACH

Some jobs in waste management require no advanced training or special skills. For example, working on a garbage collection truck is a relatively uncomplicated job. This is not to say that people holding such jobs do not perform useful work, but such work can be performed by virtually anyone who is physically mobile and can follow simple instructions. A high school diploma may be required by some employers, but even that minimum level of education is not really necessary to perform some jobs at the entry level.

For a person hoping to develop a rewarding career, such jobs may not provide enough in the way of challenges or benefits to warrant serious consideration. At the same time, they might be used as starting points for men or women ambitious enough to move up to other jobs when opportunities arise. For instance, some jobs as operators or technicians may be obtained without any postsecondary education, with on-the-job training provided to those who are hired. An aggressive person might use this as a starting point on which to build a successful career, obtaining additional training on a part-time basis while employed.

BECOMING A TRAINED TECHNICIAN

As with many other areas of employment, the waste management field generally offers more opportunity for persons with some specialized training than for those without such background. For those willing to complete such training, positions as technicians, operators, and specialists offer a viable career path. For example, specialized job categories include wastewater technician, hazardous waste specialist, and others at the technician level.

Most of these positions require special training. Sometimes this is obtained on the job, with a high school diploma as the minimum educational requirement. In other instances, training involves completion of a specific program offered by a community or technical college. This usually takes one or two years of study as a full-time student, or more on a part-time basis.

The pay off for this specialized training is that it leads to better jobs than those requiring minimum skills. Salaries tend to be greater, and job tasks often have more diversity and thus are more challenging than would otherwise be the case. In addition, technicians with the right educational credentials and job experience often enjoy the potential of advancing to supervisory positions.

WORKING AS AN ENGINEER OR SCIENTIST

For those willing to obtain substantially more education than that required for technicians, positions as engineers or scientists provide a different variety of careers in waste management and related areas.

Engineers play an important role in various aspects of waste management and recycling. They take scientific principles and apply them in a variety of practical ways. For example, an engineer might work on a large, complicated project such as designing a new wastewater treatment plant. Or a job assignment might involve something much smaller in scope, such as analyzing the efficiency of a simple pump and trying to find ways to improve its performance.

Some engineers specialize in the complex process of removing hazardous waste from a contaminated area. Others focus on recycling specific materials, development of equipment used in wastewater operations, techniques for solid waste removal, or other areas.

Engineers must be able to think analytically and to solve problems. A good grasp of basic scientific principles is generally required, as is a facility for mathematics.

A bachelor's degree is the minimum requirement for becoming an engineer today, and many engineers earn master's degrees or other advanced degrees. This requires the ability to perform well in college and the tenacity to stick with four or more years of postsecondary education.

A similar background is also required in sciences such as biology, chemistry, and geology. Careers in these fields can also be found within the general area of waste management. For example, a geologist might deal with the contamination of soil, or a biologist might deal with the use of microorganisms to break down waste materials.

Positions as scientists or engineers may offer very attractive salaries as well as providing genuinely interesting work. In addition, they often involve supervising other employees, and may be combined with, or lead to, high-level management positions.

FUNCTIONING AS A MANAGER

A technical background is not the only option for pursuing a career in waste management. An alternative is to develop skills in business administration or management and then apply them to this area.

Earning an associate, bachelor's or master's degree in a business field such as management, marketing, or accounting can potentially lead to a management position in a waste management firm or related organization. For example, a company dealing in the recycling of scrap metal may employ staff members to manage various aspects of the company's affairs. A municipal government may offer similar positions. Instead of relying on specialized technical knowledge, men and women in such positions serve as generalists who manage operations or specific business functions involved in waste management or recycling.

STARTING YOUR OWN BUSINESS

Interested in starting your own business? If so, this is another viable approach toward pursuing a career related to waste management or recycling. Many small businesses have begun operations in this area in recent years, and there is definitely room for more. For example, a small company might specialize in providing waste bins for businesses and then emptying them on a regular basis, competing with municipal waste removal services. Or a company might focus entirely on hazardous waste removal. Wherever a need exists and an individual or group of people have a marketable idea, the potential for a successful small business exists.

One of the most attractive features of such careers is that they offer a great deal of flexibility. A college degree is not necessary to start a small business, for instance, although it may be helpful. Similarly, such careers are not based so much on the decisions of others (such as whether to hire or promote you) as on your own choices and your unique talents, abilities, and ambitions.

WORKING IN RELATED AREAS

Other careers in the general area of waste management offer an assortment of possibilities. Those who develop specialized knowledge through education or experience may work as consultants, advising others on ways to solve technical problems or to achieve greater efficiency. Or they might teach in a waste management program in a two-year college. Still another possibility is working in a professional association related to the field.

Other jobs include sales of equipment or supplies to waste management or recycling companies, transporting waste products or materials related to the industry, and repairing equipment used in waste management or recycling efforts.

The following chapters provide more specific details about major categories of employment in dealing with waste materials. With the tremendous scope that this field covers, the range of job possibilities is definitely broad.

CHAPTER 3

WASTEWATER TREATMENT

A major area of employment within the overall field of waste management is wastewater treatment. This includes positions as wastewater treatment plant operators, supervisors in water treatment facilities, water purification chemists, mechanics in treatment plants, and various related jobs.

The processing of wastewater is extremely important to everyone, from individuals and families to small businesses and large corporations. In fact, it is one of the key characteristics of an advanced society. In many third world countries, the lack of an efficient system for removing waste material from water leads to serious disease or general ill health for millions of people. In South America, for example, the 1990s have seen a series of epidemics of cholera, a devastating disease that is now almost unknown in the United States and Canada. The spread of this often fatal disease is a direct result of water contamination.

Using modern technology to treat wastewater helps prevent disease, limit pollution, and protect the environment. A major purpose of wastewater treatment is to provide safe drinking water. Other purposes include limiting pollution of streams and rivers, protecting fish and other wildlife, and providing water that is safe for general use in the home and industry.

PROBLEMS WITH WATER CONTAMINATION

Water has been called the source of life, and indeed is vital to life and good health. Pure water is certainly among the most important of all substances.

Once water has been used, however, it can take on characteristics which are far from desirable. Here are some typical examples:

- Biological waste from public and private restrooms, food processing plants, medical facilities, research laboratories and other sources can enter water systems; in so doing, such waste not only contaminates the water initially, but can also serve as a medium for deadly bacteria to grow and multiply.
- Pesticides used in farming, lawn care, and general control of insects and other pests can enter ground water, lakes, streams and other sources of water. In many cases, the chemical compounds used can pose serious health dangers.

- Waste products from chemical manufacturing and other industrial processes can become water pollutants.
- Ordinary oil used in automobiles is a frequent source of water pollution. Just one quart of used motor oil, for instance, can contaminate thousands of gallons of water. In addition, used oil may contain dangerous levels of harmful chemicals.

It is the responsibility of those who work in wastewater treatment to eliminate as many pollution problems as possible. Typically, this is accomplished by maintaining wastewater treatment plants. Here, water which contains waste or pollutants is carried through sewer pipes to facilities where some or all of the following take place:

- Solid materials are removed from the water through screening, by allowing solids to settle, or through other processes;
- Microorganisms are removed through the addition of chemicals;
- Chemicals which have dissolved in the water are removed through a number of different processes;
- Oxygen is added to water by spraying it into the air or through other methods.

These and related processes require operators to move water through the facility by opening and adjusting valves, operating pumps, reading gauges and meters, and performing other duties.

EMPLOYERS

Some employers in this field include the following:

- Municipal water treatment facilities
- Investor-owned water treatment facilities
- The federal government
- State governments (and provincial governments in Canada)
- Consulting companies
- Private contractors
- Manufacturers and retailers of water treatment equipment and supplies
- Research organizations
- Educational institutions.

Because maintaining clean water benefits the entire general public, this function is often a responsibility of government. Accordingly, many workers in the field are considered public employees, since funding for their positions comes from tax dollars.

JOB TITLES

A great deal of variety can be found in the types of jobs held by persons in the general area of wastewater treatment. A sampling of job titles listed by the U.S. Department of Labor in this area includes the following:

- Basin operator
- Chemist, water purification
- Customer service representative
- Drainage-design coordinator
- Grit-removal operator
- Maintenance supervisor
- Purifying plant operator
- Sludge control operator
- Sludge filtration operator
- Superintendent, water-and-sewer systems
- Supervisor, sewer system
- Treatment plant mechanic
- Water and sewer systems supervisor
- Water purifier
- Water regulator and valve repairer
- Wastewater treatment plant attendant
- Water treatment plant mechanic
- Water treatment plant operator

WASTEWATER TREATMENT OPERATIONS

Probably the largest job category in this field is that of wastewater treatment operations. According to the U.S. Department of Labor, nearly 80,000 persons are employed in the United States alone as operators in either plants which treat water before it is used, or as wastewater treatment plant operators.

Some of the tasks performed by wastewater treatment plant operators include:

- Taking samples of wastewater;
- Analyzing water samples for chemical or biological content;
- Operating devices used to add chemicals to water;
- Reading gauges and meters used in monitoring water treatment processes;
- Adjusting equipment controls;
- Testing levels of chlorine or other purifying chemicals;
- Monitoring and repairing equipment such as pumps or valves;
- Making minor repairs to equipment;
- Using hand tools such as wrenches, pliers, and specialized tools;
- Operating computers used in controlling or analyzing water;
- Maintaining written records of treatment activities.

The *Dictionary of Occupational Titles* provides this position description for a wastewater treatment operator:

> Operates sewage treatment, sludge processing and disposal equipment in wastewater (sewage) treatment plant to control flow and processing of sewage. Monitors control panels and adjusts valves and gates manually or by remote control to regulate flow of sewage. Observes variations in operating conditions and interprets meter and gauge readings and test results to determine load requirements. Starts and stops pumps, engines, and generators to control flow of raw sewage

through filtering, settling, aeration, and sludge digestion processes. Maintains log of operations and records meter and gauge readings. Gives directions to
wastewater treatment plant attendants and sewage disposal workers in performing routine operations and
maintenance. May collect sewage sample, using dip-
per or bottle and conduct laboratory tests, using testing equipment, such as colorimeter. May operate and
maintain power generating equipment to provide
steam and electricity for plant. May be designated
according to specialized activity or stage in processing as activated-sludge operator, grit-removal operator, pump and blower operator, sludge control
operator, or sludge-filtration operator.

WORKING ENVIRONMENT

Workers in this field may experience a variety of working
conditions. Some work may take place indoors, while other
functions are performed outside. In a large treatment plant
some technicians may spend most of their time indoors,
where they are assigned to monitor specific processes or
analyze water samples; others in the same plant may spend
a great deal of time outside, where they collect water
samples, operate equipment, or perform other functions. In
a smaller plant, one worker may cover a large number of
tasks both indoors and out.

Working outside can provide both enjoyment and discomfort. Conditions may vary from beautiful spring days

to cold, dreary winter weather. Workers who encounter such situations usually learn to adjust by wearing appropriate clothing.

Indoor settings may be quite basic, especially in older buildings. More modern facilities, on the other hand, may be very comfortable. Work settings can also be noisy due to the din of machinery in operation. Workers must sometimes deal with unpleasant odors, slippery walking surfaces, and other impediments.

Most treatment plants operate around the clock, seven days a week. This means that operators usually work shifts. In some cases, an individual may work only during daytime, evening, or late night hours, following the same schedule each week. In others, employees rotate shifts on a weekly basis. Working shifts may also involve putting in time during weekends, holidays, or emergencies created by weather conditions or malfunctioning equipment.

SKILLS AND BACKGROUND NEEDED

You do not need a wide assortment of special skills to learn the work performed in wastewater operations. Anyone with good coordination, basic mechanical ability, and a willingness to learn has the potential to succeed in this field. It also helps to have many or all of these traits:

- The ability to learn basic scientific principles
- An aptitude for math

- A facility for working with hand tools
- The ability or potential to work with computers
- An interest and concern for the environment
- The capability for climbing steps and ladders, and moving about freely and easily
- Adequate eyesight for reading gauges and meters
- The ability to work independently
- The ability to follow instructions
- Reliability

LEARNING JOB SKILLS

The knowledge needed to work in wastewater operations may be obtained in two basic ways. Some employees in this field develop their skills through on-the-job training, while others complete programs offered by two-year colleges or other postsecondary schools.

Training on the Job

Many wastewater treatment operators learn their skills on the job. This usually involves the following steps.

- Earning a high school diploma or the equivalent;
- Applying for a job as an attendant or operator-in-training;
- Once hired, working under the supervision of experienced operators;

- Participating in classroom training or self-paced study
- Performing basic job tasks such as taking samples of water or sludge, observing and writing down meter readings, cleaning and servicing equipment, assisting experienced workers in carrying out routine activities

As workers become more experienced, they then become full-fledged operators or technicians. Many employers provide continuing education opportunities for such workers as they continue to gain experience, providing additional chances for increasing their understanding of the intricacies of wastewater operations.

In some cases, job openings in this field are based on results from civil service examinations, meaning that those who perform best on written examinations are offered positions. In other instances, getting a job depends as much on timing as on any particular qualifications.

Studying at a Two-Year College

An excellent way to prepare for a career in wastewater management is to enroll at a two-year college and complete a one-year or two-year program in wastewater technology, water technology, environmental engineering technology, or a related area. Many community, junior and technical colleges across the United States and Canada offer programs or courses in these areas.

For example, Green River Community College in Auburn, Washington offers an associate in applied science degree program in wastewater technology. This program can be completed by two years of full-time study. Students can also attend on a part-time basis while employed in the field or in another type of job, and take three years or more to earn a degree. To be admitted, students must have taken high school chemistry and algebra, or be able to show the equivalent through standardized test results or college courses taken prior to enrollment in this program.

Green River's program prepares students for jobs as wastewater technicians working in wastewater collection or treatment facilities. It includes several general college courses such as English and chemistry, and a larger number of courses dealing specifically with job-related skills.

For students who do not wish to earn an associate degree but focus instead on vocational skills, an option is available which leads to a vocational training certificate. This requires students to complete 58 credits on the quarter system (about one calendar year of full time study) compared to 95 credits for an associate degree.

Students pursuing a vocational studies certificate complete the following courses:

- Water Hydraulics
- Pump, Valve and Piping Systems
- Utility Work Safety
- Chemical Feed Systems
- Water and Wastewater Applied Problems
- Water and Wastewater Laboratory
- Drawings and Manuals

- Instrument and Control Systems
- Wastewater Distribution
- Wastewater Treatment 1
- Wastewater Treatment 2
- Introduction to General Chemistry
- Written Communication or Introductory Communication
- Technical Skills
- Elementary Algebra
- Introduction to Microcomputers

To earn an associate degree, students complete the same courses needed for a certificate, plus the following:

- Field Experience Preparation
- Wastewater Field Experience or Cooperative Education
- General Biology (or Natural Science and the Environment)
- General Chemistry (two courses)
- Fundamentals of Oral Communication
- Freshman English

The courses dealing specifically with wastewater technology cover the basics needed for employment in the wastewater treatment industry following program completion.

For instance, students studying "Wastewater Collection" become familiar with the following:

- Composition and sources of wastewater;
- Purposes of wastewater collection systems;

- Components of wastewater systems including preliminary treatment, piping systems and lift stations;
- Wastewater collection systems operation and maintenance including safety procedures, inspecting, testing, repair and cleaning;
- Normal and abnormal operating conditions.

The college's courses in ''Wastewater Treatment 1'' and ''Wastewater Treatment 2'' cover the following:

- Introduction to wastewater treatment technology;
- Purposes of wastewater treatment;
- Purposes, operation, and maintenance of wastewater treatment plant components;
- Pre-treatment and primary treatment;
- Trickling filters and rotating biological contactors;
- Purposes and operation of oxidation ditches, waste treatment ponds, and activated sludge;
- Disinfection;
- Sludge digestion;
- Solids handling and effluent disposal.

Another course, ''Wastewater Field Experience,'' provides supervised field experience for students. Operating as trainees in actual wastewater collection or treatment systems, students learn to perform routine operation or maintenance procedures. This can be an excellent supplement to classroom and laboratory instruction.

A similar option is available to students who sign up for Cooperative Education field experiences. Here, students not only gain up to 9 quarter hour credits, but also earn salaries or wages from the facilities where they serve.

Because of the specialized nature of wastewater technology, it may not be offered as an academic program in a community or technical college near your home. To find out, consult catalogs or contact any school's admissions office.

Some community and technical colleges combine courses in wastewater technology with those dealing with the general supply of water. Others offer either a certificate or associate degree program, but not both.

Waubonsee Community College in Sugar Grove, Illinois, offers degree and certificate programs in what it calls "environmental control technology." These programs prepare students to work as water treatment plant operators, wastewater treatment plant operators, sanitarians, and environmental health inspectors.

Courses offered in this program include:

- Wastewater Treatment I
- Wastewater Treatment II
- Water Treatment I
- Chemical Analysis of Water and Wastewater
- Water Treatment II
- Water and Wastewater Instrumentation and Equipment Fundamentals
- Atmospheric Pollution Control
- Wastewater III—Process Control.

After completing a certificate (requiring completion of 24 semester hours) or an associate degree (64 semester hours), students at Waubonsee are prepared to seek jobs with government agencies, private industries, private water supply companies, chemical manufacturing companies, or the armed forces.

Many two-year colleges which do not offer full-scale programs in wastewater technology have at least some courses which can provide background in this area. Programs in environmental engineering technology, in particular, are becoming increasingly common. Even a course or two in this or related programs can provide useful preparation for a career in this area. A partial listing of colleges offering such courses can be found in Appendix E and F.

Non-credit workshops and seminars can also provide useful training, especially for persons who are already employed in the field. The American Water Works Association, for example, provides a variety of workshops on topics of interest to persons employed in the industry. Most states and provinces also offer courses and seminars through their water pollution control agencies. Many training manuals and other materials are also available to the individual who is aggressive enough to take advantage of them.

A certificate or two-year degree may not always be necessary to gain entry into the general field of wastewater treatment and operations. As previously noted, in some cases a high school diploma is the minimum. But as technology becomes more complex, special training of this type is becoming increasingly important. Even in situations where a degree or certificate is not a requirement, possessing one is sure to enhance anyone's prospects for landing a job, or for advancing within the hiring organization as opportunities for promotions develop.

RELATED CAREER POSSIBILITIES

In addition to positions as operators and attendants or operator-trainees, a number of other career paths can be followed in the area of waste treatment and operations. Some of these are as follows:

- Chemists
- Engineers
- Laboratory technicians
- Operator supervisors
- General laborers or helpers
- Clerical personnel
- Superintendents

The position of operator-supervisor represents a logical goal for a person who wants to advance on the job after serving as an operator. The duties for such a job include the following tasks noted by the *Dictionary of Occupational Titles:*

> Supervises and coordinates the activities of workers engaged in operating and maintaining equipment in wastewater (sewage) treatment and disposal facility.

> Directs activities of workers in power generating, grit
> removal, pump and blower, and sludge filtration de-
> partments. Tests sewage and water supply to prevent
> contamination, using chemicals and testing equip-
> ment. Inspects equipment, such as pumps, sediment
> tanks, filters and chlorinators to detect malfunctions.
> Confers with engineering personnel and directs work-
> ers in repair and maintenance of machinery and equip-
> ment. May direct workers in maintenance of buildings
> and grounds.

Men or women who advance to the position of operator-
supervisor may potentially move even farther. The highest
position in a typical treatment plant is that of superintendent. This job consists of providing overall direction for
all activities conducted by the organization. It requires
very good management skills, and in return offers not just
a good salary, but also the challenge of leading an entire
organization.

QUESTIONS TO ASK YOURSELF

If a career in this area seems of interest, it may be useful
to ask yourself questions such as these:

1. Have you taken courses at the high school level in
   biology or chemistry, or are you willing to take them?
2. Do you enjoy working with your hands?
3. Are you confident in working with tools?

4. Are you concerned about the well-being of the environment?
5. Are you comfortable with math?
6. Can you picture yourself working in a wastewater treatment plant or a similar facility?
7. Do you have good eyesight (or can problems be corrected with glasses or contact lenses)?
8. Have you finished high school, or are on track to high school completion?
9. Are you willing to pursue additional education if necessary?

If you can answer "yes" to most or all of these questions, then a career in wastewater operations may be a good match with your own background, abilities, and potential.

CHAPTER 4

SOLID WASTE MANAGEMENT AND RECYCLING

One of the most fascinating aspects of waste management is the scope of challenges caused by the accumulation of waste materials. Consider, for example, materials which are routinely discarded by your own immediate family. One day's trash may include items such as the following:

- Milk cartons made of plastic or coated paper
- Glass jars
- Newspapers or magazines
- Food scraps
- Metal cans
- Dirt removed from floors
- Discarded clothing
- Plastic or paper shopping bags
- Spent batteries
- Broken glass
- Used aluminum foil
- Aerosol cans

Now multiply one day's total by seven to get an idea of the volume produced in a week by your family, and then again by fifty-two for an entire year's worth of garbage. On top of that, think of other items discarded over the course of a year: used tires, broken electronic equipment, half-empty paint cans, old furniture, and more. Then multiply the cumulative volume by the number of households in your city, town or rural area. The resulting estimate should begin to provide just one indication of the volume of waste generated in our society.

The example noted above applies only to waste material generated by private residences. An entirely different dimension is added when the practices of business and industry are included. Industrial wastes such as ash produced in the burning of coal, slag generated in making iron and steel, fibers produced in chemical and textile industries, and many by-products add millions of tons of waste to the environment each year.

Consider these examples of waste products generated in the United States alone, based on figures compiled by the Environmental Protection Agency (EPA):

- Each year, Americans throw away more than 2 million tons of rubber tires;
- In a single month, over a million tons of newspapers are discarded;
- More than 10 million tons of old television sets, radios, toasters and other electronic devices and components are thrown away annually;

- A year's accumulation of discarded shoes and clothing totals more than 4 million tons;
- Over 13 million tons of food waste are thrown away each year;
- Paper, plastics, yard wastes, textiles, wood, glass and other materials are discarded in huge quantities.

All told, people in the United States generate more than 160 million tons of trash each year. As pointed out by the EPA, this would fill a convoy of 10–ton garbage trucks 145,000 miles long—or enough distance to circle the equator almost six times!

The production of wastes presents a range of challenges. How could the millions of discarded tires, for instance, be put to some useful purpose? How can we deal with toxic wastes produced by the chemical industry? What should be done with radioactive wastes produced by nuclear power plants? How can we most effectively make use of the potential offered by recycling? Such questions will continue to generate interest in waste management issues for years to come.

The enormity of this field also means that it generates thousands of jobs. Large-scale production of wastes means that large numbers of people are needed to collect, transport, store, destroy or recycle the millions of tons of solid waste produced every year in the United States and Canada. As a result, this represents a promising career area.

TYPES OF JOBS

A variety of jobs can be found within the general area of solid waste management, as well as in recycling (discussed later in this chapter). The U.S. Department of Labor lists a number of job titles in the area of solid waste management, including the following:

- Sanitary landfill operator
- Sanitary landfill supervisor
- Disposal plant operator
- Incinerator plant—general supervisor
- Solid waste facility operator
- Solid waste facility supervisor
- Garbage collection supervisor

The most basic jobs involve physical labor, such as collecting refuse from private homes and businesses on a daily route or accepting and burying garbage brought to a landfill. More rewarding careers can be found in positions which supervise these functions, or in related technical areas such as engineering.

The Dictionary of Occupational Titles includes these duties in the job description for a sanitary landfill operator:

> Performs any combination of following duties to dispose of solid waste materials at landfill site. Operates heavy equipment, such as bulldozer, front-end loader, and compactor to excavate landfill site, transport solid waste materials, and to spread and compact

layers of waste and earth cover. Directs incoming vehicles to dumping area. Examines cargo to prohibit disposal of caustic waste, according to government regulations. Weighs vehicles entering and leaving site and collects dumping fees.

WORKING ENVIRONMENT

By the very nature of their jobs, many landfill operators, and those holding related positions, must spend a substantial amount of time outdoors. If they are operating a piece of earthmoving equipment to bury garbage, for instance, they may be in an enclosed cab, but they will still find themselves frequently exposed to the elements. This may be seen as something of an asset to those who do not enjoy being confined indoors. It can also be a source of discomfort, especially in Canada and many parts of the United States where cold weather conditions—including snow, sleet and subfreezing temperatures—are not uncommon.

Workers in this area must also deal at times with mud, unpleasant odors, and generally unsanitary working conditions. They must pay special attention to job safety, since movement of garbage and soil can be dangerous. Most workers wear gloves, safety shoes and other protective equipment as needed.

BACKGROUND AND SKILLS NEEDED

Few special skills are needed to work in solid waste management. Anyone with good physical fitness and a willingness to learn the required job tasks could consider this career area.

It would be helpful to have at least some of the following traits before pursuing work in this area:

1. Good driving ability
2. Physical strength (or at a minimum, above average physical endurance)
3. An aptitude for working with tools or heavy equipment
4. A willingness to work in an environment which may not be attractive or comfortable
5. A strong work ethic
6. A cooperative work attitude.

WHERE EMPLOYMENT IS

Men and women are employed in solid waste management positions throughout the United States and Canada. Positions can be found in both rural and urban areas, since waste products are generated virtually everywhere.

Employers include sanitation departments of cities, towns and states as well as private companies.

BREAKING IN

For many jobs in solid waste management, you do not need to attend college to break into the field. A high school diploma may be required, however.

Some vocational schools at the high school level teach basic skills which can be applied to the work performed in this field. Learning to operate or repair any type of equipment, for example, might show employers that you have potential for working with trucks, end-loaders, or other heavy equipment.

Courses offered by two-year or four-year colleges in environmental technology, civil engineering, engineering technology or related areas can be valuable, particularly if you aspire to a supervisory position. The same is true of business courses if you hope to become a manager in an area related to solid waste management.

Many positions in this field require no preliminary training. Skills are learned on the job. Applicants for jobs demonstrate that they have the potential to become good workers or that they have developed good work habits elsewhere, and then employers provide training on the job for those who are hired.

RELATED POSITIONS

Related positions include jobs as supervisors and managers in solid waste facilities, as well as persons operating their own businesses. Technicians and engineers in special applications such as recycling and hazardous waste management (discussed in the next chapter) also can find employment in this field.

Supervisors and managers in solid waste management need familiarity with the basic operations involved as well as the ability to perform management tasks such as:

- Supervising daily activities of workers;
- Assigning employee work schedules;
- Planning work flow and identifying activities to be accomplished;
- Evaluating the job performance of subordinates;
- Completing written reports;
- Reporting job problems and accomplishments to higher level supervisors or owners;
- Insuring that safety standards are met;
- Monitoring expenditures for equipment, supplies and other materials.

RECYCLING CAREERS

In addition to jobs involved in basic aspects of solid waste management, a number of career opportunities can be found dealing with the recycling of waste products. In fact, recycling offers some of the most exciting prospects in this field.

Interest in recycling has increased dramatically in recent years, as more attention has been placed on conserving our natural resources. A survey conducted for *McClean's* magazine showed that two-thirds of Canadians feel recycling should be required by law. A number of the United States have already passed laws requiring that certain materials be recycled.

Currently, about 10 percent of all trash is recycled, according to the Environmental Protection Agency (EPA), with a much higher proportion expected in the future. The EPA cites the following advantages of recycling:

1. It reduces the nation's reliance on incinerators and landfills.
2. Recycling is less expensive than incineration or landfilling.
3. It reduces the amount of harmful substances released into the environment.
4. The practice of recycling reduces the need for raw materials and thus conserves natural resources.

Recycled materials are used for a variety of purposes. Here are some examples cited by the EPA:

- Glass can be refilled or used to make new bottles, jars, and other glass items.
- Plastic drink bottles and other items made of plastic can be reprocessed to make auto parts, fiberfill, and other useful products.

- Aluminum cans are recycled and used to make sheet aluminum and castings.
- Yard waste can be composted for landscaping.
- Animal waste is used as fertilizer.
- Clothing and furnishings can be reused by other people.
- Paper is reconstituted and used again as newsprint, cardboard, or insulation, among other purposes.

Businesses, government agencies, schools, churches, families, volunteer groups, and many others are becoming increasingly interested in recycling. This means that greater quantities of materials are being recycled. As a result, more jobs than ever before can be found within the area of recycling.

Recycling Iron and Steel

One of the largest areas for recycling involves iron and steel, also known as ferrous products. This is one part of the recycling industry that has existed for quite some time. The necessary technology has been in place for decades, and the economic advantages of reusing iron and steel have long been recognized.

The basic concept behind recycling ferrous products is simple. When iron or steel is melted, it can be reshaped and used for virtually any purpose slated for such metal when it was originally formed. While in the molten state, other materials such as iron oxide (rust), paint, nonferrous metals, and nonmetallic materials can be efficiently removed. Of course melting iron and steel requires tremendously high temperatures and special equipment, but the same needs also apply to production of ferrous materials in the first place. Thus the technology for recycling iron and steel is already well established.

An old car, for instance, may contain a ton or more of iron and steel which can be reused for other purposes. Other products such as steel beams, farm equipment, household appliances, tools, and even toys can serve as sources of material for recycling.

The Institute of Scrap Recycling Industries, an association of companies dealing in scrap commodities, has reported that its member companies helped recycle more than 58 million tons of iron and steel in 1990 alone. This included over 9 million automobiles. According to the association, such a total would be enough to manufacture 580 million new refrigerators!

The large volume of material involved means that a significant number of jobs can be found related to the recycling of iron and steel. Employers of persons working in this field include:

- Small companies which collect or buy scrap metal from individuals, companies, government organizations, farms and other sources, and then sell it to larger recycling companies;

- Small, medium, and large companies which not only buy scrap metal, but also prepare it for recycling by disassembling and packing it;
- Brokers who act as go-betweens in buying and selling scrap iron and steel.

These and other companies employ men and women to perform a wide variety of functions, from collecting and sorting steel, to operating large machines such as cranes or huge shredders used to tear apart autos or other machinery.

Nonferrous Metals

In addition to iron and steel, a number of other metals can be recycled. In its publication "Recycling Non-Ferrous Scrap Metals," the Institute of Scrap Recycling Industries lists the following metals which are most commonly recycled:

Aluminum is widely used for purposes ranging from aircraft and automobile components to soft drink cans. Nearly 3 million tons of aluminum are recycled yearly in the United States.

Copper is used in electrical wiring, electronic components, plumbing materials and in alloys with other metals, among other purposes.

Lead is recycled in large quantities. A primary use is in automobile batteries as well as smaller batteries used in a variety of applications.

Zinc's uses include forming metal alloys, galvanizing steel, and manufacturing paint.

Nickel is used in forming stainless steel and other metal alloys.

Other metals which are recycled, according to the institute, include:

Tin
Cobalt
Chromium
Tungsten
Magnesium
Titanium
Manganese
Zirconium
Silver
Gold
Platinum
Columbium

Recycling of these metals increases the availability of resources which exist in finite quantities. It also saves energy, because recycling of metals generally consumes much less electricity or other energy than does the processing of original ores. In addition, the reuse of metal products reduces the volume of solid waste that must be handled and stored. The increasingly common practice of recycling soft drink cans, for instance, removes millions of cans from the environment.

While the reclaiming of aluminum cans is a highly visible example, it represents only a portion of overall efforts to recycle metals of various types. In many cases, metal collected for recycling is not even a used material, but rather a by-product of manufacturing. Known as ''new scrap'' or ''industrial scrap,'' this material must be recycled before it can be of any practical use. For example, molten brass may be poured into molds to form bookends, paperweights, or other decorative ornaments; when the molds are broken, extra pieces of brass are left over. The same is true of many other manufacturing processes. Defective or broken items also provide similar material for recycling. Because it is unused, this type of scrap typically is free of many of the contaminants found in used metals, making it ideal for recycling.

Regardless of the source of the material, the recycling of nonferrous metals provides a variety of job opportunities. Employers include scrap dealers and processors; intermediate users such as smelters or refining firms; and end users such as foundries, mills, and various types of manufacturers. Jobs range from yard laborers and operators of special equipment used in crushing or cutting metals, to technicians and engineers who develop new types of metal products from recycled materials.

Paper Recycling

The reuse of paper provides additional career opportunities, with job opportunities increasing as paper recycling becomes more commonplace.

The demand for paper products has never been greater. Where a few years ago some people were predicting that electronic communications would largely replace paper as a medium for exchanging information, just the opposite has occurred. Computers, facsimile (fax) machines, and printing presses spew out printed information at previously unheard of rates, and more paper is being used than ever before. Add to that the continued use of paper for cardboard boxes, paper towels, packing, and other products, and the importance of paper is undeniable.

According to the Paper Stock Institute, over 75 million tons of paper are used in the United States every year. Of this total, about 20 million tons are recycled.

When paper is recycled, trees are saved and energy used in production is reduced. Recycled paper is commonly used to produce a variety of products including newsprint, corrugated boxes, paper towels and tissues, stationery, and many other products.

Jobs based on the recycling of paper range from the collection of used paper to the production of items made from paper which has been recycled. Persons employed in this area work as laborers, equipment operators, technicians, chemists, supervisors, and other positions.

FUTURE OF RECYCLING

As recycling grows increasingly popular, more and more products and materials will be reused instead of discarded or destroyed. The United States has established a national goal of having 25 percent of all trash recycled, and this goal may not only be reached, but exceeded.

In years ahead, recycling is destined to become increasingly important. After all, more and more of the earth's resources are being used up. Many of them cannot be replenished. Growing use of recycling practices will be a virtual necessity.

From a career perspective, this is good news. Whether you are interested in working for a company that specializes in recycling or starting your own recycling business, this is certainly an area of enormous potential.

CHAPTER 5

HAZARDOUS WASTE MANAGEMENT

In recent years a whole new industry has come into existence within the overall field of waste management: hazardous waste management. This area offers a growing number of career opportunities.

Until the second half of the twentieth century, waste material was generally seen as more of a nuisance than a danger. There was little widespread understanding of the chemical and physical properties of waste materials, and of the potential dangers some materials hold toward human health. But technological developments, industrial accidents, legislative actions, and growing public concern about waste-related hazards have opened up a new area in dealing with waste materials. Some of these developments are described below.

RADIOACTIVE WASTES

The development of nuclear technology has been accompanied by a complex question: how can radioactive waste materials be disposed of safely? Before World War II this was virtually an unknown issue, but the successful invention of atomic weapons in 1945 led to further developments of a broad scope. Not only have nuclear weapons been produced by the thousands since that time, but a follow-up to weapons development was the harnessing of nuclear energy for peaceful purposes. Nuclear power plants are now common in industrialized countries, and they provide an extremely efficient source of electricity.

A problem with both the production of nuclear weapons and the creation of atomic energy, however, is that radioactive materials are produced either as by-products or as leftover waste. Unfortunately, such materials emit harmful radiation which can cause illness or death. To compound matters, nuclear wastes can remain radioactive for hundreds of years, meaning they pose long-term hazards as well as short-term dangers. If radioactive materials enter the atmosphere, soil, or water supply, they can lead to radiation poisoning, genetic mutations, cancer, and other catastrophic problems.

TOXIC CHEMICALS

As modern society has advanced, an area of tremendous change has been the creation of thousands of new chemical compounds, as well as the development of many new chemical and industrial processes which create various by-products. Many of these products and by-products are quite beneficial. Plastics, pharmaceutical drugs, petroleum products, herbicides, cosmetics, food additives, and synthetic materials used in clothing are just some examples.

Some chemicals, however, pose definite threats to human beings. One example is dioxin. This chemical was originally developed for the useful purpose of killing weeds and thereby helping farmers do a more efficient job of producing food, among other purposes. After extensive use of this chemical, however, it was discovered that dioxin can cause serious health problems in animals and people. Furthermore, it does not readily break down into other compounds, meaning it can spread through the environment and continue to present problems long after the initial point of contamination.

Dioxin is just one of thousands of toxic substances. Many other chemicals have been found to be dangerous to good health. Some are useful for specific purposes, but dangerous if not handled or discarded properly. Others represent waste materials left over from chemical reactions or industrial processes. All too often, these wastes were buried in old drums which eventually began to leak, or merely dumped on the ground or in the nearest body of water. After years of being forgotten or ignored, such sites began to generate concern when the health of people living near them began to suffer.

Probably the most dramatic example of hazardous waste problems has been the Love Canal incident in the state of New York, where an entire town had to be permanently evacuated due to uncontrolled disposal of industrial waste products. But this was only one of many such catastrophes. According to the Environmental Protection Agency (EPA), more than half a million American waste dumps hold significant amounts of toxic substances.

To address this problem, the U.S. Congress and various state and provincial governments have adopted laws designed to limit the dumping of potentially harmful wastes. They have also allocated funds for the process of cleaning up such wastes. For instance, the government has allocated billions of dollars for a huge ''Superfund'' expressly for this purpose. Administered by the EPA, this program focuses on the most dangerous waste sites. The massive scale of this effort not only indicates how important the issue of hazardous waste control has become, but it also means that thousands of new jobs have been created in this area. Similar programs sponsored by other government agencies have also created many new jobs in the United States and Canada.

MEDICAL WASTES

The disposal of medical wastes represents another area where potential health hazards exist. This has long been a matter of importance, but concerns about such risks have been multiplied in recent years by the AIDS epidemic and all the publicity which has accompanied it.

A typical hospital must dispose of all of the following:

- Blood samples
- Tissue samples
- Diseased organs removed in operations
- Used syringes
- Used bandages
- Other types of medical wastes

With the increased attention given to safe disposal of such materials, new opportunities are being created for firms and individuals who can perform such services.

TYPES OF JOBS

Many of the positions discussed in previous chapters may involve working with hazardous wastes. Operators in solid waste landfills, for instance, may be responsible for disposing of toxic substances. Similarly, the wastewater treatment process may include dealing with dangerous chemicals. In addition, a number of jobs are based largely on working with hazardous wastes.

According to the U.S. Department of Labor as reported in *Occupational Outlook Quarterly,* some jobs which may deal directly with hazardous wastes include the following:

- Analytical chemists
- Biologists
- Chemical engineers
- Civil engineers
- Environmental chemists
- Environmental engineers
- Environmental protection specialists
- Environmental technicians
- Engineering geologists
- Geochemists
- Industrial hygienists
- Hydrologists
- Organic chemists
- Risk management specialists
- Soils engineers
- Spill engineers
- Toxicologists
- Wastewater engineers

WORK PERFORMED

The work performed by persons in these various jobs is in many ways comparable to that undertaken in waste management positions described in other chapters. Engineers specializing in this area, for example, tend to have similar credentials and hold responsibilities not terribly unlike those of engineers working in other waste management fields. The major difference is their focus on eliminating the hazards of toxic wastes or other dangerous substances.

Most job responsibilities in this field fall under these two broad areas: 1. removal of dangerous materials; and 2. remediation or ''fixing'' of contamination and environmental damage.

Removal of Dangerous Materials

The practice of removal deals with the elimination of immediate hazards by a toxic waste spill, industrial accident, or similar event. Such happenings are not uncommon; in fact, thousands of these incidents happen every year. For example, say a train is derailed while traveling through a suburban area, and in the process a tanker car filled with hydrochloric acid turns over, thereby releasing thousands of liters of acid. A quick response to this problem is needed. Specially trained technicians—perhaps employed by a local government, the company which produced the acid, or a private company specializing in hazardous waste removal—are called to the scene. Here they take initial steps to neutralize the acid, and then remove the remaining material as efficiently and completely as possible. This process might take hours or days, and might involve any number of workers.

In some instances, dealing with a waste spill can be relatively simple, with the chemicals in question simply scooped or suctioned in containers and hauled away, or perhaps treated with another chemical and rendered harmless. In other instances, the work is much more complex. For example, large quantities of contaminated soil might have to be removed in a difficult and time-consuming process.

Remedial Actions

In contrast to the immediacy of an industrial or transportation accident, remedial actions tend to focus on cleaning up waste that has accumulated over a period of time. A typical example might be a dump which has been used by a chemical company to dispose of drums containing various chemical wastes, and which has now been targeted by the Environmental Protection Agency for clean-up efforts. Work involved in this remediation process would include determining what kinds of chemicals are present, removing drums and disposing of them through incineration or some other effective method, and then cleansing the soil and ground water of any substances that have been spilled or leaked.

JOB CATEGORIES

Generally, workers in this field fall within these areas:

- Technicians, specialists and other support personnel;
- Engineers;

- Chemists, biologists, geologists and other scientific personnel;
- Managers and administrative/clerical personnel.

Technicians and other support staff conduct much of the hands-on work involved in hazardous waste management. Specific tasks undertaken by such workers might include the following:

- Collecting samples of waste materials;
- Assisting in laboratory analysis of waste products;
- Removing toxic substances from contaminated sites;
- Recommending waste removal action plans to engineers or managers;
- Driving vehicles used in waste removal;
- Maintaining and repairing equipment ranging from pumps to end-loaders;
- Providing general assistance to engineering or scientific personnel.

Engineers, chemists and other highly trained workers may perform functions such as the following:

- Planning or conducting investigations of waste sites;
- Identifying toxic substances;
- Studying the feasibility of ''cleaning up'' waste sites;
- Developing specific strategies for waste removal and remediation;
- Supervising technicians and other workers;
- Completing various types of written reports;
- Evaluating the effectiveness of waste removal or remediation efforts.

Managers, administrative support staff, and clerical workers in the hazardous waste field tend to perform a variety of functions not unlike those of managers in other settings. Many combine technical backgrounds with training or experience in business administration or management. In some cases, they are entrepreneurs who develop their own businesses to specialize in hazardous waste management.

TYPICAL EMPLOYERS

Employers of persons in this field include the following:

- The U.S. government
- State and provincial governments
- Chemical manufacturing firms
- Companies specializing in hazardous waste removal

A typical company in the last category might deal in any or all of the following:

- Asbestos abatement (including site assessment, removal and disposal);
- Lead abatement;
- Removal of heavy metals;
- Removal or clean-up of petroleum products or chemical solvents;
- Assessment and clean-up of chemical spills;

- Removal and disposal of storage tanks;
- Clean-up of toxic waste dumps.

BACKGROUND AND TRAINING NEEDED

The diverse jobs in this area involve different training requirements. Jobs as technicians or specialists generally require at least a high school diploma, with on-the-job training sometimes available. Those who complete a one- or two-year program in a community or technical college would have an ideal background for many such positions, especially if they receive a degree or certificate in environmental studies, waste management, or other related area.

For engineering or scientific positions, at least a bachelor's or master's degree is necessary. In addition, course work dealing with hazardous substances will be an asset. Many colleges and universities offer one or more courses in this area which can be taken by students who are majoring in engineering, chemistry, geology, or biology. In addition, a number of schools offer special programs or courses in environmental studies, hazardous waste management or a related field.

The ultimate in preparation for entering this field is the completion of an advanced degree in environmental science or a related area. Such programs can be completed at a number of institutions, including New Jersey's Rutgers University, which offers graduate degrees in environmental sciences. Areas in which master's degrees or doctorates can be earned include:

- Treatment and disposal of environmental pollutants;
- Environmental contamination and analysis;
- Air, industrial hygiene, and environmental health;
- Soil science;
- Radiation science (doctoral only);
- Exposure assessment (doctoral only, offered in cooperation with other institutions).

Students enrolled in these programs may pursue specializations in areas such as the following:

- Air pollution and resources
- Aquatic biology
- Aquatic chemistry
- Aquatic toxicology
- Chemistry and physics of aerosol and hydrosol systems
- Environmental chemistry
- Environmental biology
- Ground and surface water modeling
- Human exposure
- Occupational health
- Pesticides residue chemistry
- Radiation biology
- Radiation chemistry
- Radiation physics
- Risk assessment
- Soil mineralogy and micromorphology

- Soil chemistry
- Soil microbiology
- Soil physics
- Solid and hazardous wastes
- Water and wastewater treatment, and
- Water resources.

Completion of a bachelor's degree in engineering or a scientific discipline is required before a student can be admitted to any of these programs, including background in biology, chemistry, physics, and math. A master's degree can be completed in one or two years of full-time study, while a doctorate normally takes several years beyond the bachelor's or master's degree.

A graduate degree in environmental science can lead to a job in research, teaching, or direct field work (such as a toxic waste removal and clean-up project supported by funding from the Environmental Protection Agency).

While a majority of colleges and universities do not offer programs as specialized as those found at Rutgers, many offer specializations or individual courses dealing with environmental management, hazardous wastes, or related areas. A degree in chemistry, biology, engineering, or a related discipline can be enhanced by completion of one or more courses of this type, and may qualify degree recipients for positions in hazardous waste management, solid waste management or other similar areas.

CHAPTER 6

EDUCATION AND TRAINING OPTIONS

You may need to complete a special training or educational program if you want to pursue a career in waste management or recycling. This may consist of on-the-job training, courses offered by a two-year college, or courses offered by a four-year college or university.

For some positions, such as garbage collection crews, no special training is required. Many of the better-paying and more challenging jobs, on the other hand, are based on special knowledge which must be gained through some type of formal training.

DEVELOPING EDUCATIONAL PLANS

Before taking any action toward career preparation, take some time to develop specific plans regarding your possible education. Planning ahead can help prevent you from wasting time, money, or both. In this process, consider questions such as the following:

- What type of education is required to qualify for the type of career in which you are interested?
- Is on-the-job training a possibility?
- If formal education at the college level is required, what types of programs are available? Is a two-year college or a four-year college the best choice?
- What schools offer waste management or related programs in your geographical area?
- Are you willing or able to go away to college, or do you need to stay close to home?
- Are you willing to go to college for two years, four years or more?
- How difficult is the course work?
- Do you have the necessary background to succeed? (For example, if you are interested in engineering, do you have an adequate background in math?) If not, are you prepared to complete remedial courses?
- How much will completion of an educational program cost?
- How will you pay for tuition and other educational costs?
- What steps should you take to make certain that you succeed?

Give careful consideration to such questions before choosing a specific career direction or enrolling in an academic program. A little advance planning will help assure that your choices are the right ones for your own needs and interests.

ON-THE-JOB TRAINING

Some employers offer their own training programs or contract with consultants or staff from professional associations to provide such training. Under these programs, short-term classes may be offered in a format similar to that in schools. Or new workers may be given a few informal instructions and then work under the supervision of more experienced workers, learning as they go about daily tasks.

To find out about employer-sponsored training programs, check with the personnel office at any company or organization that employs waste management personnel. If such an option exists and you are interested in participating, file an application for employment and indicate that you are willing and eager to complete any required training. This can be a good opportunity to break into the field.

If you are selected for on-the-job training, try to follow these steps to make sure you succeed:

1. Always be on time. Don't make a bad impression by failing to get to work or class on time.
2. Be an active participant. As new information is presented, make sure you understand it. Ask questions. Show that you are genuinely interested in learning as much as you can.

3. Try hard. Give any learning experience your best effort. Learn from mistakes. If you have a positive attitude, good results are likely. Remember, the first days and weeks on any new job will be crucial in showing employers you have the potential to be an excellent worker.
4. Be respectful to supervisors, training staff and more experienced workers. In most cases, these people will be older than you. Even if not, their experience and knowledge deserve respect.

COLLEGE PROGRAMS

For many careers in waste management, the best place to start is a college-level program in the field. If you plan to work as an operator or technician, a community or technical college may be your best choice. For an engineering or scientific degree, you will need to attend a four-year college or university.

A look at higher education options follows. For details about specific programs and courses, see the previous chapters.

Community, Junior and Technical Colleges

For preparation in wastewater operations, environmental technology, business management and a number of related areas, a viable educational option is attending a junior, community, or technical college. Many of these two-year colleges offer a choice between a basic program which can be completed in a year or less, or an associate degree program which normally takes two years as a full-time student to complete.

Associate degree programs take longer to finish because students must take classes not just in waste management or other area of primary interest, but also in other subjects. For instance you might also study psychology, history, English composition or technical writing. Relatively few such classes must be completed, but they are required before you can earn a two-year college degree.

It is important to realize that many courses completed at two-year colleges can be transferred to four-year colleges and universities. This may or may not be in your plans at first, but educational goals often change. Earning credits which can be transferred may be to your advantage in the future.

If you want to complete your studies in as short a time as possible, another option is to pursue a community college program which leads to a diploma or certificate instead of a degree. With this option, courses may not be accepted as transfer credits at four-year schools, neither will you need to take courses in general studies. Instead, you will study only courses within your major area of interest.

Four Year Colleges and Universities

If you want to become an engineer or scientist, it will be necessary to attend a four-year college or university. It may be possible to spend the first two years at a community college and then transfer, but eventually attendance at a four-year institution will be needed.

Many universities and four-year colleges offer programs in various engineering fields. Even if a program in a waste management field is not offered, a traditional engineering or science degree may be of interest. In some cases, this can serve as a foundation for students who later earn an advanced degree in a specialized area related to waste management.

Some colleges emphasize nontechnical areas such as the liberal arts, and do not offer any engineering programs at all. To find out if a program is available at any given college or university, consult the college catalog or the admissions office. If you are interested in programs beyond the bachelor's degree, check with the graduate school at any university in which you are potentially interested.

Most four-year schools, unlike community colleges, do not practice open admissions. Students must meet certain admission requirements, and many colleges operate on a competitive basis and accept only some applicants. Before applying to a four-year college, be sure to find out about the admissions process—what kind of information is required and when it must be submitted.

SELECTING A COLLEGE

In choosing a college, take some time to check out basic details about the school. What kind of institution is it? Take a close look at the school's catalog and see whether students earn diplomas, certificates, associate degrees, or bachelor's degrees. Are advanced degrees awarded? Also consider information such as:

- The kinds of jobs graduates are prepared to perform
- How many courses must be completed
- Length of time to complete a program
- Descriptions of courses related to waste management
- Accreditation (Beware if none is listed. Colleges should be accredited by a regional accrediting group such as the North Central Association of Colleges and Schools, the Southern Association of Colleges and Schools or the New England Association.)

If possible, visit the campus. Enrolling sight unseen can lead to problems. Take a look at labs, classrooms, and other facilities. Make sure the environment is one in which you can feel comfortable.

In talking with school officials or others who are familiar with the institution (such as counselors), don't be reluctant to ask questions. What is the school's placement rate for those who have graduated? How is it viewed in the community? How do former students feel about it? Does it have a good reputation? Ask questions such as these whenever possible.

Also, analyze costs and your ability to pay them. Tuition, fees, and other expenses vary widely from one school to another. The least expensive schools are usually public institutions funded through tax revenue. Those assessing the lowest costs for attending tend to be public two-year colleges. Most community, junior, and technical colleges attempt to keep costs as low as possible so that almost anyone can afford to attend. Many two-year colleges charge less than $1,200 for tuition and fees for a full academic year. This is very reasonable when compared with the tuition charged by most four-year colleges.

Private colleges usually charge substantially more than public institutions since they do not receive government support. This can be offset to some degree by the fact that their students may receive larger amounts of student aid than those at public institutions.

College costs can include some or all of the following:

- Tuition (the basic cost of classes);
- Fees (may be a synonym for tuition, or may apply to other costs);
- Application fees (often required before enrollment, and usually nonrefundable even if you decide not to attend);
- Book costs (these are not usually charged along with tuition, but represent an extra cost students must pay, usually to the school's bookstore; may be several hundred dollars for a single term);

- Lab fees (often charged to help cover the cost of equipment and supplies);
- Activity fees (may be charged even for students who do not participate in recreational or cultural activities);
- Health fees (special fees may be charged to support student health services for all students);
- Room and board (may be charged directly by the school, or may consist of costs to live off-campus);
- Commuting expenses (for students who commute, costs can include gasoline, bus costs, car upkeep and other expenses);
- Other fees (various other fees may be charged such as parking fees, costs for taking special tests, having transcripts sent to employers or other schools, dropping or adding classes, and other purposes).

While the greatest expense is for tuition and basic fees, it is important not to overlook other costs in making educational plans.

GETTING FINANCIAL AID

Any way you look at it, college is expensive. Still, students who need financial help can usually obtain it. If money is a problem, consider applying for financial aid. This can come in the form of a grant, scholarship, loan, work-study award or other form of financial assistance.

For many students, the best source of aid is the U.S. government. Millions of students receive money from the government every year through a variety of financial aid programs. Other programs for loans, grants, or scholarships are sponsored at the state level.

To obtain student aid through most government programs, you must show financial need. The needier you are, the more funding you can expect. Students who need only some degree of help can qualify for special loans offered at low interest rates.

Other sources of aid include schools and colleges themselves. Many offer a variety of financial aid awards including scholarships, loans, and grants. Thousands of private organizations also sponsor special aid programs ranging from scholarships to grants.

In obtaining financial aid, the key is to be aggressive in pursuing assistance. For students willing to fill out forms, meet deadlines, and provide the needed information, chances of receiving financial aid are excellent.

Applying for Federal Aid

To obtain student aid from the government, you must first provide information (which remains confidential) about your family's income, assets, debts, and other financial matters. This is done by completing a detailed application form.

Any of the following forms can be completed to initiate a request for student aid:

- The U.S. Department of Education's Application for Federal Aid (AFSA)
- The American College Testing Program's Family Financial Statement (FFS)
- The College Scholarship Service's Financial Aid Form (FAF)
- The Student Aid Application for California (SAAC)
- The Pennsylvania Higher Education Assistance Agency's Application for Pennsylvania State Grant and Federal Student Aid

You can get these forms from the sponsoring agency, from high school guidance counselors, and from financial aid offices in colleges and trade schools. If you are unsure which form is best for your situation, check with a counselor or financial aid officer.

Be certain to meet application deadlines. The best time to apply is around January 1 of the calendar year in which you plan to begin your postsecondary studies. Some schools may suggest an even earlier time period. Since some federal programs award money on a "first come, first served" basis, the earlier an application is submitted, the better. At any rate, make certain you apply before May 1.

Major Aid Programs

Several types of aid programs are offered by the government. Following is a brief overview of major awards available. Keep in mind that many students receive a "package" of aid consisting of several different types of awards.

PELL GRANTS

These grants are designed for students who have genuine financial need. Many consider them the most desirable type of award available. After all, a grant does not have to be repaid!

The amount any one student receives will vary according to individual finances, costs at the school being attended, and related factors. Recent students have received anywhere from a few hundred dollars to well over $2,000 yearly, and the upper limit may be raised to an even higher amount. This award is based on need—not grades or other academic factors.

If you really need a Pell Grant, make certain to apply properly. Financial limitations should not keep you from pursuing a college education.

SUPPLEMENTAL EDUCATIONAL OPPORTUNITY GRANTS

(SEOG) awards are much like Pell Grants, but not as many awards are available each year. They, too, need not be repaid. Since overall funds are limited, it is very important to apply early if you hope to land one of these grants.

LOAN OPTIONS

Loan programs provide another set of options. Some are offered by the government itself, while others come from private lending agencies with government backing. They must be repaid after your education is completed, but most offer a long time to repay as well as lower interest rates than ordinary commercial loans.

One of the most popular loan programs is the Perkins Loan program. This provides loans with relatively low interest rates and plenty of time to repay the loan. Another widely used program is the Stafford Loan program, which offers similar benefits. The latter is different, however, in that loans are obtained directly from a bank, credit union, or other financial institution. Interest rates are lower than conventional loans thanks to government backing of the loans.

Other loans are available which place less emphasis on financial need. Many families have incomes which are too high for need-based programs, but a loan would still be helpful in meeting educational expenses. For such situations, PLUS (Parent Loans for Undergraduate Students) loans or SLS loans (Supplemental Loans for Students) offer an attractive alternative. PLUS loans are made directly to parents of students, while SLS funds are taken out by students. Both types of loans are made through banks or other private lenders, and the major requirement is a good credit rating.

WORK-STUDY PROGRAM

Many students earn money through the College Work-Study Program, where students hold part-time jobs at their college or a cooperating agency. Typical jobs include working in a dean's office, running the switchboard, serving as a lab assistant, or helping out in the college's library, bookstore, cafeterias, or ground crew.

Participants in this program earn at least the federal minimum wage. They also gain job experience. This can be helpful in showing prospective employers that they have been successful workers, as well as providing the chance to receive letters of recommendation from college staff members who have served as work-study supervisors.

OTHER SOURCES OF AID

In addition to aid programs offered by the government, other sources of assistance are available. For example, consider possibilities such as these:

- Scholarships and grants offered by individual schools
- Grants or scholarships offered by professional associations
- Scholarships sponsored by organizations to which you or a parent belong, such as churches or civic clubs
- Tuition programs sponsored by companies for their employees (a great possibility if you can gain employment before completing an educational program, or wish to move up from one level to another).

CHAPTER 7

SALARIES, WAGES AND BENEFITS

One of the most attractive features of a career in waste management or recycling is that it can provide a good income, as well as a number of fringe benefits. Because of the diversity of jobs in this field, actual salaries vary widely. In general, though, men and women employed in this field enjoy good earnings potential, especially if they hold jobs which require special training.

EARNINGS

At the low end of the scale, entry-level workers in jobs requiring few special skills earn relatively modest wages. The lowest amount is generally $4.25 per hour, the federal minimum wage, which equals an annual salary of between $8,000 and $9,000. In some cases wages at the entry level are significantly higher, depending on factors such as location and whether a union contract is in effect. Wages of

$5.00 to $10.00 per hour are not uncommon for jobs such as working on a garbage collection crew or serving as an operator trainee in a wastewater treatment plant.

Highly trained workers such as engineers, on the other hand, can earn excellent salaries. Most engineers earn well above the average for all jobs, according to the U.S. Department of Labor, with annual salaries of $50,000 or more not unusual.

Supervisors and business owners also can earn excellent salaries; in fact, the potential is virtually unlimited in the latter category. An owner of a successful business dealing with waste management or recycling could earn even more than engineers or scientists.

Here are some typical examples of wages and salaries offered for positions in this field:

- According to the American Water Works Association, entry-level salaries for laboratory and field technicians average $17,000 to $23,000 yearly.
- Salaries for wastewater treatment plant operators average $20,000 to $30,000, according to the U.S. Department of Labor. The salary scale for such a position employed by the city of Asheville, North Carolina, was $14,622 to $26,437 in 1991, as noted in a classified ad in *Waterworld News*.
- The position of assistant water and sewer superintendent in Bloomfield Township, Minnesota, was recently offered at a salary of $42,000 to $44,000 a year in an

ad listed in the *American Water Works Association Journal.*

- The Virginia Department of Waste Management employs men and women as environmental program managers at salaries of $32,910 to $44,952 per year, according to state salary schedules.

As these figures show, salaries vary widely depending on the type of work done as well as a number of other factors. Some of these additional factors which help determine salaries and wages include the following:

Location Employees in urban areas usually are paid more than those in rural areas, since urban areas tend to have a higher cost of living. For instance, food and housing are generally more costly in Chicago or Montreal than in rural Alabama. As a result, workers in waste management as well as other fields can expect correspondingly higher pay.

Education In most cases, jobs requiring a greater level of educational preparation pay higher salaries. An engineer with a bachelor's or master's degree, for instance, will usually earn more than a technician with two years or less of post-high school training, or a worker with only a high school diploma.

Related experience Generally, workers with experience in their field are paid more than those with limited experience. Within a given organization, those who have been employed the longest tend to earn the highest salaries. Similarly, a person who takes a new job should command a higher starting salary than one who is just starting his or her career.

Economic conditions The overall state of the local or national economy also has an effect on wages and salaries. During good economic times, salaries and wages usually increase. In periods of recession, salaries may experience limited growth. In some cases, such as when a company is experiencing a financial crisis or when public budgets are cut, they may even be reduced.

Supply and demand If employers have a difficult time finding trained workers, they will generally pay more to get them. Thus anyone with special skills or training in waste management will have an advantage if there are more jobs available than persons to fill them. On the other hand, if unemployment rates are high or there is a surplus of workers in a given specialty area, it is unrealistic to expect salaries to be any higher than the minimum employers normally pay.

Type of employer Employers vary a great deal in their ability to pay competitive salaries as well as their overall approach to this issue. A small, family-run scrap business may keep salaries as low as possible simply to insure survival of the company. A public water treatment plant may follow salary schedules established by local or state government officials. If a union is in place, wage levels will be established through collective bargaining, and may be higher than would otherwise be the case. The size, financial status and management philosophy of any employing organization will have a great deal of impact on the level of salaries and wages.

FRINGE BENEFITS

Employers provide various fringe benefits to their employees along with normal salaries or wages. These benefits may vary substantially from one employer to another. They are usually more extensive for full-time employees than for those working on a part-time basis.

Fringe benefits are important. They help employees plan for retirement, deal with illnesses, and enjoy holidays or vacations. In some cases, they are worth more in a practical sense than a significant portion of regular salary paid. A person earning $20,000 a year with no benefits, for example, may not be as well off as one making $15,000 yearly who also receives a generous package of fringe benefits.

Employees in waste management and recycling may receive any or all of the following benefits:

- Paid vacations
- Pension/retirement funds
- Paid sick leave
- Medical insurance
- Worker's compensation in case of injury
- Social Security benefits
- Profit-sharing plans
- Dental insurance

- Optical insurance
- Life insurance
- Performance bonuses.

Before accepting an offer of employment, it is important to learn what benefits are available. Any decision to accept a position should include an analysis of benefits as well as regular salary or wage potential.

CHAPTER 8

ORGANIZATIONS AND CERTIFICATIONS

Serving as a technician, engineer, operator, manager or other worker in the waste management field involves many challenges. A number of organizations help people meet such challenges by providing support for individual workers, companies, or both. Such groups include labor unions, professional societies, and other organizations.

LABOR UNIONS

Some workers in waste management or recycling jobs belong to labor unions. These organizations promote the welfare of workers in relations with employers. Unions have been an important part of American and Canadian society for more than one hundred years, and in that time they have helped achieve major advancements such as shorter work weeks, better pay and fringe benefits, and recognition of basic rights of workers.

Typical rights and benefits of labor union membership include the following:

- Basic membership rights to any worker who wishes to join the union;
- The right to attend union meetings and participate in discussions about topics of interest to the membership;
- The right to vote to elect officers of the organization;
- Representation in collective bargaining between the union and the employer;
- Protection from arbitrary or unfair treatment by employers;
- Strike benefits for members of the organization if a strike occurs;
- Special benefits such as retirement funds and group rates for life insurance.

Unions are primarily financed by dues from members. A typical amount paid by one member is about two hours' wages per month. In return for this investment, members benefit from the union's efforts to attain better wages, working conditions and employee rights for its members, as well as the other advantages listed above.

Union leaders negotiate with employers to seek benefits such as cost-of-living wage increases, paid sick leave, pension benefits, and safe working conditions. They also work to influence legislation on matters such as worker rights, tax reform, government safety regulations, and trade policies.

Whether or not any one worker belongs to a union depends on a number of factors, including the following:

Type of employer. The waste management industry is not as heavily unionized as some others. Many scrap recovery businesses, for instance, are family owned and operate on a nonunion basis. However, employees of large firms such as steel companies, and those employed by large city governments, often do belong to unions.

Location. Unions are more common in some geographical locations than in others. In the southern United States, they are not a dominant factor in industry. In the industrialized northeast United States and parts of Canada, labor unions are more commonplace.

Type of position. Even within a unionized setting, all employees may not belong to a union. Most managers, for instance, do not hold union membership. Operators, technicians and other workers, on the other hand, are more likely to belong.

PROFESSIONAL SOCIETIES

Many persons interested in waste management, recycling, or related areas (whether or not they belong to unions) also may be involved in professional societies designed to provide specific advantages to members as well as to advance the field in general. Following is an overview of several of these organizations. A listing of addresses for selected waste management organizations is included in Appendix B.

American Water Works Association

The American Water Works Association (AWWA) is the major organization for persons employed in wastewater management and related facets of the water industry. Based in Denver, Colorado, this group provides a wide range of services to members. In addition, it furnishes information to Congress and other governmental bodies on issues of importance to the industry.

Both organizational and individual memberships are available. Members may take advantage of the following benefits and services:

- Information on legislation such as the Safe Drinking Water Act (SDWA);
- Educational programs including seminars, workshops and courses;
- Publications about the water industry;
- Training and other services related to prevention of on-the-job accidents;
- Assistance in complying with government regulations;
- Public relations materials.

In addition, individual members can choose from services such as these:

- Job listings and other employment services;
- Health insurance at reduced rates through a group plan;
- Group life insurance plans;
- Discounts on publications;
- Awards and recognition for professional accomplishments;
- Membership in a local chapter of the organization;
- Discounts on regular fees for seminars, workshops and conferences;
- The right to display the association's logo on stationery, business cards and other materials;
- Industry information, including both printed publications and access to a computerized data base.

Publications offered by the association include a monthly journal which provides news, technical reports, and other information; a monthly newsletter dealing with water treatment plant operations; a newspaper, also published monthly, which covers news and activities about association and industry activities; and a bimonthly tabloid covering developments of a technological nature. In addition, the association publishes an annual buyers' guide about goods and services related to water supply and treatment; a publications catalog; and a variety of materials and books about the water industry.

For students enrolled in colleges and universities, a special student membership is available. Members can also choose from *active* or *affiliate* memberships.

The association's employment services can be particularly helpful for students and other persons just beginning their careers. Through published job listings, for example, vacancies can be located in wastewater operations and other areas.

Institute of Scrap Recycling Industries

The Institute of Scrap Recycling Industries (ISRI) is a trade association based in Washington, D.C. Its members include about 1,800 companies which engage in the business of recycling scrap materials.

The ISRI was formed in 1987 when two long-standing organizations merged: the National Association of Recycling Industries (formed in 1913) and the Institute of Scrap Iron and Steel (formed in 1928). Members include both small and large businesses. Many are family owned with a long tradition in the industry.

Services provided by this organization include the following:

- Information for the public about recycling and scrap processing;
- Government relations activities to promote industry needs and viewpoints with governmental representatives;
- Training activities for persons working in scrap processing;

- Information on safety in the workplace;
- Public relations programs encouraging recycling.

The institute has 23 chapters which provide members with a local or regional link. The organization also employs a professional staff which manages institute activities.

National Solid Wastes Management Association

An important trade group serving both the United States and Canada is the National Solid Wastes Management Association (NSWMA). This 2,500 member group is headquartered in Washington, D.C., and operates 31 state and provincial chapters. Members represent various aspects of the waste management industry including sanitary landfill operations, rubbish collection and disposal, hazardous and medical waste treatment and disposal, recycling, and resource recovery.

The association sponsors institutes which address topics such as the following:

- Rubbish collection
- Landfill operation
- Recycling
- Biomedical waste treatment
- Nonbiological hazardous waste treatment and disposal
- Waste-to-energy plant operation
- Equipment manufacturing and distribution.

Other activities of NSWMA include:

- Monitoring the development of policies, regulations and laws related to waste management;
- Communicating needs and opinions of the association and its members to lawmakers and other government officials;
- Collecting and distributing information about waste management issues to members as well as to the public;
- Offering seminars and conferences on waste management topics;
- Sponsoring a convention and trade show attended by more than 5,500 persons annually.

The organization also publishes a magazine, *Waste Age,* and a biweekly newspaper, *Recycling Times.* While maintaining a number of field offices which help address local and regional issues, the main focus of the organization is on national waste management issues.

Solid Waste Association of North America

The Solid Waste Association of North America (SWANA) provides a variety of educational and professional development services. Most members are persons actively working in waste management positions.

The organization maintains chapters throughout Canada and the United States. These chapters place particular emphasis on waste management issues such as the following:

- Integrated planning for municipal solid waste management;
- Collection, disposal, and transfer of municipal solid waste;
- Hazardous waste management;
- Landfill operation;
- Procurement policies and practices;
- Markets for recycling;
- Conversion of waste to energy;
- Legislative issues addressed by state, provincial and national government bodies.

This association offers a number of continuing education opportunities as well as special certification programs. It also publishes and distributes publications.

A strong point of the association's work is the technical assistance it provides. This includes a special "Peer Match" program, through which members may receive technical help from others in the field. Under this program, individuals who have questions or need technical advice are matched with colleagues who can provide the necessary advice or assistance. Members also can obtain help from association staff members through site visits, written advice or telephone conversations.

The association also sponsors a computerized data base on various topics related to solid waste management. Known as SWICH (Solid Waste Information Clearinghouse), the system includes a library system of reports, studies, films, videotapes, and other materials. It also features an electronic bulletin board covering solid waste issues, and includes a toll-free telephone number. Topics covered by SWICH include the following:

- Source reduction
- Composting
- Planning
- Recycling
- Education and training
- Legislation and regulation
- Public participation
- Collection
- Waste combustion
- Disposal
- Transfer
- Landfill wastes
- Special wastes

Other Organizations

In addition to those organizations discussed here, other groups can prove useful in meeting specific needs. Those who become instructors in waste management fields, for example, may find it beneficial to participate in the American Vocational Association, the American Education Association, or statewide groups of professional educators. Persons who start their own businesses in this field, or who work in management positions, may want to participate in organizations such as the American Management Association or other business-related organizations.

CERTIFICATIONS

In some waste management positions, it is desirable to undergo a special certification process to demonstrate work-related competencies. In fact, this may be a requirement. Wastewater treatment plant operators, for example, are required in 46 states to pass a special certification exam, according to the U.S. Department of Labor. The few states which do not require certification offer voluntary programs of this type.

Professional organizations also provide certification opportunities. The Solid Waste Association of North America, for instance, offers a special "Professional Certification Program" (PCP) for managers, technicians, and designers of municipal solid waste management systems. It also sponsors a training and certification program for managers of municipal solid waste landfills and for landfill enforcement officers.

The certification process provides benefits both to the individual and to the employer. It also helps ensure the public welfare in making certain that workers in this important area are competent. Advantages of the certification process include:

- Safety and health issues are emphasized;
- Employers can gain confidence in workers having demonstrated competency in standard areas;
- Educators and trainers can have a "target" for use in planning courses and programs;
- Individual workers can increase their own self-confidence through successful completion of certification exams;
- Public health is fostered;
- An overall sense of professionalism is enhanced for men and women employed in the field.

Succeeding with Certification Exams

Before taking an exam for certification purposes, be sure to prepare for the test by gaining as much technical information as possible. If you have not completed a college-level program, take advantage of any seminars or workshops which may be offered by your company or a professional organization. Also, read through manuals or other background materials.

In addition, try to follow the following steps as you complete the examination process:

1. Read all written questions before answering any of them. Then go back and start answering each one. This will give you an idea of the overall nature of the questions, while allowing your mind to begin ad-

dressing more than one question at the same time, at least on a subconscious basis.

2. If time limits are imposed, be sure to keep track of them. Avoid spending too much time on any one question. It is a good idea to wear a watch, since a clock may not be provided.

3. Study ahead of time. This helps in two ways. First, your overall knowledge will be increased, which of course is the primary consideration. Second, your confidence will be higher and you will be less likely to suffer from nervousness.

4. Before sitting for the exam, try to talk with someone who has already completed it (such as a co-worker or instructor). Ask for advice about the experience, and then use it during your own preparation.

5. During the exam, don't be reluctant to pose questions to the person administering the test. You cannot ask for help, but if you have any questions about procedures for completing the examination, bring them up.

6. Try to control your nerves. Taking any test can be an imposing experience. Try to limit nervousness by talking with friends before the exam, or by using calming techniques such as deep breathing. If you have taken the time to prepare, you should not be overly worried.

Once you have passed any certification exam, make the most of it. Be certain your supervisor knows that you have succeeded, as well as anyone else who might be interested (such as former teachers). Also, be sure to keep records related to certification. In the future, you may need to show proof of that certification to obtain a new job or a promotion.

CHAPTER 9

TAKING THINGS FROM HERE

Does a career in waste management or recycling seem challenging now that you have reviewed the material provided in previous chapters? If so, future actions on your part might include any (but probably not all) of the following:

1. Determining what kinds of training programs are available in the areas in which you are interested;
2. Applying for a summer job in a waste management field to gain exposure to the type of work involved;
3. Writing to colleges and requesting catalogs and information about any programs they might offer in areas related to waste management;
4. Applying for admission to a two-year college;
5. Applying for admission to a four-year college or university;
6. Applying for financial aid;
7. Completing a diploma or degree program in waste management or a related field;

8. Landing an entry-level job and completing an on-the-job training program;
9. Joining a professional organization related to waste management;
10. Developing plans to start your own company specializing in recycling or some other aspect of waste management;
11. Talking with teachers, counselors, or people already employed in the field about prospects for a waste management career

NEGOTIATING THE JOB SEARCH PROCESS

After finishing a college program, or whenever you feel prepared to look for a position, you will need to seize the initiative in seeking employment. This will involve identifying job openings, submitting applications and participating in job interviews.

Identifying Job Openings

Many job openings are listed in the classified sections of newspapers. A Sunday edition is often the best in terms of numbers of openings listed. Here is a newspaper ad for a challenging position with the Virginia Department of Waste Management:

Environmental Program Manager

The Department of Waste Management is seeking an individual to assume responsibility for management of statewide groundwater monitoring system. Incumbent will be responsible for developing technical guidelines for evaluating groundwater portion of permits/closure plans; assessing technical/regulatory merit of systems; developing/maintaining project work schedules and supervising work of and training new staff; evaluating adverse environmental impact of facilities on groundwater; conducting compliance meetings with facility owners/operators. Knowledge of solid waste management practices/theory and regulatory standards required. Knowledge of construction principles and basic sciences necessary; knowledge of project management desirable. Demonstrated written and oral communication abilities are required. Bachelor's degree in engineering or earth sciences preferred; experience in waste management required; supervisory and project management experience preferred. Hiring salary range: $32,910 to $44,952.

Professional journals are another good source of job listings. A recent issue of the *American Water Works Association Journal,* for example, listed job vacancies for:

- Environmental managers
- Environmental engineers
- Sales manager for a water pump company
- Assistant water and sewer superintendent
- Civil engineer for water/wastewater operation.

Some companies post job announcements on bulletin boards in addition to (or in place of) newspaper advertising. Company personnel offices will also have such information.

Your local employment service or job service office (provided by state or local government as a free service) also receives listings of job openings.

Your college's placement office is another source of job information. Most colleges operate such offices with the specific purpose of finding jobs for their students and graduates.

Getting a Job

After identifying a job opening, obtain and complete a job application. In the process, take your time in filling it out, and answer all questions completely and honestly. Type the information or write as neatly as possible, and doublecheck spelling and grammar.

If you are invited for a job interview, make the most of the opportunity by taking steps such as these:

Be prepared Before the interview takes place, plan ahead for the experience. Try to anticipate possible questions and practice answering them. Also, try to obtain some basic information about the employer so you will be better prepared to ask intelligent questions.

Be neat Even if the position doesn't require you to wear dress clothes, do not make the mistake of showing up for an interview looking sloppy. First impressions really do count. Be sure to wear clean, neat clothes and aim for a well-groomed appearance.

Arrive on time Always be on time for an interview. Failure to show up on time may make an employer wonder if you will have a problem reporting to work on time if hired.

Control your nerves Although interviews can be imposing, try to stay calm. Keep in mind that if you do not land this job, there will be other interviews.

Sooner or later, that initial job in waste management or recycling should be yours. Your career can then move forward, potentially leading to many years of satisfying and genuinely important work.

APPENDIX A

BIBLIOGRAPHY

"Bibliography of Municipal Solid Waste Management Alternatives." United States Environmental Protection Agency, 1989.

"Characterization of Municipal Solid Waste in the United States: 1990 Update." United States Environmental Protection Agency, 1990.

"Meeting the Challenges." American Water Works Association. (not dated)

"Public Attitudes Toward Garbage Disposal." National Solid Wastes Management Association, 1990.

"Recycling Non-Ferrous Scrap Metals." Institute of Scrap Recycling Industries, 1990.

"Recycling Paper." Institute of Scrap Recycling Industries, 1990.

"Recycling Scrap Iron and Steel." Institute of Scrap Recycling Industries, 1990.

"Recycling Works." United States Environmental Protection Agency, 1989.

Stanton, Michael. "Hazardous Wastes: Who's Cleaning Up." *Occupational Outlook Quarterly.* Winter, 1987, 3–14.

"Scrap: America's Ready Resource." Institute of Scrap Recycling Industries. (not dated)

"The Solid Waste Dilemma: An Agenda for Action." United States Environmental Protection Agency, 1989.

U.S. Department of Labor. *Dictionary of Occupational Titles,* 1987.

U.S. Department of Labor. *Occupational Outlook Handbook,* 1990.

SELECTED WASTE MANAGEMENT ORGANIZATIONS

American Water Works Association
  666 West Quincy Avenue
  Denver, Colorado 80235

Institute of Scrap Recycling Industries
  1627 K Street, NW
  Washington, DC 20006

National Solid Wastes Management Association
  1730 Rhode Island Avenue, NW
  Suite 1000
  Washington, DC 20036

United States Environmental Protection Agency
  401 M Street, SW
  Washington, DC 20460

Solid Waste Association of North America
  P.O. Box 7219
  Silver Spring, MD 20910

APPENDIX C

STATE RECYCLING OFFICES

Following is a list of state recycling offices:

Alabama

Department of Environmental Management
  Solid Waste Division
  1715 Congressman Wm. Dickinson Drive
  Montgomery, AL 36130

Alaska

Department of Environmental Conservation
  Solid Waste Program
  P.O. Box O
  Juneau, AK 99811–1800

Arizona

Department of Environmental Quality–O.W.P.
  Waste Planning Section, 4th Floor
  Phoenix, AZ 85004

Arkansas

Department of Pollution Control and Ecology
    Solid Waste Division
    8001 National Drive
    Little Rock, AR 72219

California

Recycling Division
    Department of Conservation
    819 19th Street
    Sacramento, CA 95814

Colorado

Department of Health
    4210 E. 11th Avenue
    Denver, CO 80220

Connecticut

Recycling Program
    Department of Environmental Protection
    Hartford, CT 06106

Delaware

Department of Natural Resources and Environmental Control
    89 Kings Highway
    P.O. Box 1401
    Dover, DE 19903

District of Columbia

Public Space and Maintenance Administration
    4701 Shepard Parkway, S.W.
    Washington, DC 20032

Florida

Department of Environmental Regulation
2600 Blairstone Road
Tallahassee, FL 32201

Georgia

Department of Community Affairs
40 Marietta St., N.W., 8th Floor
Atlanta, GA 30303

Hawaii

Litter Control Office
Department of Health
205 Koula Street
Honolulu, HI 96813

Idaho

Department of Environmental Quality
Hazardous Materials Bureau
450 W. State Street
Boise, ID 83720

Illinois

Illinois EPA
Land Pollution Control Division
2200 Churchill Road
P.O. Box 19276
Springfield, IL 62706

Indiana

Office of Solid and Hazardous Waste Management
Department of Environmental Management
105 S. Meridian Street
Indianapolis, IN 46225

Iowa

Department of Natural Resources
   Waste Management Division
   Wallace State Office Building
   Des Moines, IA 50319

Kansas

Bureau of Waste Management
   Department of Health and Environment
   Topeka, KS 66620

Kentucky

Resources Management Branch
   Division of Waste Management
   18 Reilly Road
   Frankfort, KY 40601

Louisiana

Department of Environmental Quality
   P.O. Box 44307
   Baton Rouge, LA 70804

Maine

Office of Waste Reduction and Recycling
   Department of Economic and Community Development
   State House Station #130
   Augusta, ME 04333

Maryland

Department of Environment
   Hazardous and Solid Waste Administration
   2500 Broening Highway
   Building 40
   Baltimore, MD 21224

Massachusetts

Division of Solid Waste Management
    D.E.Q.E
    1 Winter Street, 4th Floor
    Boston, MA 02108

Michigan

Waste Management Division
    Department of Natural Resources
    P.O. Box 30028
    Lansing, MI 48909

Minnesota

Pollution Control Agency
    520 Lafayette Road
    St. Paul, MN 55155

Mississippi

Non-Hazardous Waste Section
    Bureau of Pollution Control
    Department of Natural Resources
    P.O. Box 10385
    Jackson, MS 39209

Missouri

Department of Natural Resources
    P.O. Box 176
    Jefferson City, MO 65102

Montana

Solid Waste Program
    Department of Health and Environmental Science
    Cogswell Building, Room B201
    Helena, MT 59620

Nebraska

Litter Reduction and Recycling Programs
 Department of Environmental Control
 P.O. Box 98922
 Lincoln, NE 68509

Nevada

Energy Extension Service
 Office of Community Service
 1100 S. Williams Street
 Carson City, 89710

New Hampshire

Waste Management Division
 Department of Environmental Services
 6 Hazen Drive
 Concord, NH 03301

New Jersey

Office of Recycling
 Department of Environmental Protection
 CN 414
 401 E. State Street
 Trenton, NJ 08625

New Mexico

Solid Waste Section
 Environmental Improvement Division
 1190 St. Francis Drive
 Sante Fe, NM 87503

New York

Bureau of Waste Reduction and Recycling
Department of Environmental Conservation
50 Wolf Road, Room 208
Albany, NY 12233

North Carolina

Solid Waste Management Branch
Department of Human Resources
P.O. Box 2091
Raleigh, NC 27602

North Dakota

Division of Waste Management
Department of Health
1200 Missouri Avenue, Room 302
Box 5520
Bismarck, ND 58502–5520

Ohio

Division of Litter Prevention and Recycling
Ohio EPA
Fountain Square Building, E-1
Columbus, OH 43224

Oklahoma

Solid Waste Division
Department of Health
1000 N.E. 10th Street
Oklahoma City, OK 73152

Oregon

Department of Environmental Quality
  811 S.W. Sixth
  Portland, OR 97204

Pennsylvania

Waste Reduction and Recycling Section
  Division of Waste Minimization and Planning
  Department of Environmental Resources
  P.O. Box 2063
  Harrisburg, PA 17120

Rhode Island

Office of Environmental Coordination
  Department of Environmental Management
  83 Park Street
  Providence, RI 02903

South Carolina

Department of Health and Environmental Control
  2600 Bull Street
  Columbia, SC 29201

South Dakota

Energy Office
  217-1/2 West Missouri
  Pierre, SD 57501

Tennessee

Department of Public Health
  Division of Solid Waste Management
  Customs House, 4th Floor
  701 Broadway
  Nashville, TN 37219–5403

Texas

Division of Solid Waste Management
   Department of Health
   1100 W. 49th Street
   Austin, TX 78756

Utah

Bureau of Solid and Hazardous Waste
   Department of Environmental Health
   P.O. Box 16690
   Salt Lake City, UT 84116–0690

Vermont

Agency of National Resources
   103 S. Main Street, West Building
   Waterbury, VT 05676

APPENDIX D

PUBLICATIONS AVAILABLE FROM THE ENVIRONMENTAL PROTECTION AGENCY

The following publications are available at no charge from the EPA RCRA/Superfund Hotline. Call (800) 424–9346.

General

America's War on Waste–Environmental Fact Sheet	EPA/530-SW-90-002
Bibliography of Municipal Solid Waste Management Alternatives	EPA/530-SW-89-055
Characterization of Municipal Combustion Ash, Ash Extracts, and Leachates– Executive Summary	EPA/530-SW-90-029B
Characterization of Municipal Solid Waste in the United States: 1990 Update– Executive Summary	EPA/530-SW-90-042A

Decision-Maker's Guide to Solid Waste Management (Volume I)	EPA/530-SW-89-072
Reusable News–Winter	EPA/530-SW-90-018
Reusable News–Spring	EPA/530-SW-90-039
Reusable News–Summer	EPA/530-SW-90-055
Sites for Our Solid Waste: A Guidebook for Effective Public Involvement	EPA/530-SW-90-019
Siting Our Solid Waste: Making Public Involvement Work	EPA/530-SW-90-020
Unit Pricing: Providing an Incentive to Reduce Waste	EPA/530-SW-91-005
Variable Rates in Solid Waste: Handbook for Solid Waste Officials–Executive Summary	EPA/530-SW-90-084A

Source Reduction

Be an Environmentally Alert Consumer	EPA/530-SW-90-034A
Characterization of Products Containing Lead and Cadmium in Municipal Solid Waste in the United States, 1970 to 2000– Executive Summary	EPA/530-SW-89-015C
The Environmental Consumer's Handbook	EPA/530-SW-90-034B

Recycling

| Recycling Brochure | EPA/530-SW-88-050 |
| Recycling Works! | EPA/530-SW-89-014 |

Used Oil

How to Set Up a Local Program to Recycle Used Oil	EPA/530-SW-89-039A
Recycling Used Oil: For Service Stations and Other Vehicle Service Facilities	EPA/530-SW-89-039D
Recycling Used Oil: 10 Steps to Change Your Oil	EPA/530-SW-89-039C
Recycling Used Oil: What Can You Do	EPA/530-SW-89-039B

Plastics

Methods to Manage and Control Plastic Wastes–Executive Summary	EPA/530-SW-89-051A
The Facts About Plastics in the Marine Environment	EPA/530-SW-90-017B
The Facts on Degradable Plastics	EPA/530-SW-90-017D
The Facts on Recycling Plastics	EPA/530-SW-90-017E
Plastics: The Facts About Production, Use, and Disposal	EPA/530-SW-90-017A
Plastics: The Facts on Source Reduction	EPA/530-SW-90-017C

Educational Materials

Adventures of the Garbage Gremlin	EPA/530-SW-90-024
Let's Reduce and Recycle: Curriculum for Solid Waste Awareness	EPA/530-SW-90-005
Recycle Today: Educational Materials for Grades K–12	EPA/530-SW-90-025

| Ride the Wave of the Future: Recycle Today! | EPA/530-SW-90-010 |
| School Recycling Programs: A Handbook for Educators | EPA/530-SW-90-023 |

The following EPA publications are available for a fee from the National Technical Information Services (NTIS). Call (703) 487–4650.

Characterization of Municipal Solid Waste Combustion Ash, Ash Extracts, and Leachate	PB90-187 154
Characterization of Municipal Solid Waste in the United States: 1990 Update	PB90-215 112
The Effects of Weight- or Volume-Based Pricing on Solid Waste Management	PB91-111 484
Methods to Manage and Control Plastic Wastes	PB90-163 106
Office Paper Recycling: An Implementation Manual	PB90-199 431
Promoting Source Reduction and Recycling in the Marketplace	PB90-163 122
Variable Rates in Solid Waste: Handbook for Solid Waste Officials	PB90-272 063
Yard Waste Composting: A Study of Eight Programs	B90-163 114

FOUR-YEAR COLLEGES AND UNIVERSITIES OFFERING PROGRAMS RELATED TO WASTE MANAGEMENT

Many four-year colleges offer programs related to waste management. These include programs in environmental engineering, ecology, and many other related academic disciplines as noted in earlier chapters.

Following is a list of just some of the institutions which place emphasis on this area. To find out if any given college or university offers programs related to waste management, request a catalog and peruse the listing of programs and courses offered. Or contact the admissions office for undergraduate programs, or the graduate school for programs beyond the bachelor's degree level.

Alabama

Auburn University
  Auburn University, AL 36849

California

Humboldt State University
  Arcata, CA 95521

University of California–Santa Barbara
  Santa Barbara, CA 93106

University of California–Los Angeles
  Los Angeles, CA 90024

University of Southern California
  Los Angeles, CA 90089

Florida

Florida Institute of Technology
  Melbourne, FL 32901

Florida International University
  Miami, FL 33199

University of Central Florida
  Orlando, FL 32816

University of South Florida
  Tampa, FL 33620

Illinois

Northwestern University
  Evanston, IL 60208

Southern Illinois University
  Carbondale, IL 62901

University of Illinois
  Urbana, IL 61801

Indiana

Purdue University
  West Lafayette, IN 47907

Iowa

University of Iowa
Iowa City, IA 52242

Kansas

Kansas State University
Manhattan, KS 66506

Kentucky

University of Louisville
Louisville, KY 40292

Thomas More College
Crestview Hills, KY 41017

Louisiana

Southern University–Baton Rouge
Baton Rouge, LA 70813

Maryland

The Johns Hopkins University
Baltimore, MD 21218

Massachusetts

Massachusetts Institute of Technology
Cambridge, MA 02139

Tufts University
Sommerville, MA 02144

Michigan

Michigan Technological University
Houghton, MI 49931

University of Michigan–Ann Arbor
   Ann Arbor, MI 48109

Western Michigan University
   Kalamazoo, MI 49008

Mississippi

Mississippi Valley State University
   Itta Bena, MS 38941

New Jersey

Stevens Institute of Technology
   Hoboken, NJ 07030

Rutgers University
   New Brunswick, NJ 08903

Nebraska

University of Nebraska–Omaha
   Omaha, NE 68182

New Mexico

New Mexico Institute of Mining and Technology
   Socorro, NM 87801

New York

Clarkson University
   Potsdam, NY 13676

Cornell University
   Ithaca, NY 14853

Rensselaer Polytechnic Institute
   Troy, NY 12180

State University of New York at Stony Brook
   Stony Brook, NY 11794

Syracuse University
   Syracuse, NY 13244

North Carolina

Duke University
   Durham, NC 27706

Oklahoma

Oklahoma State University
   Stillwater, OK 74078

University of Oklahoma
   Norman, OK 73019

Oregon

Oregon State University
   Corvallis, OR 97331

Pennsylvania

California University of Pennsylvania
   California, PA 15419

Carnegie Mellon University
   Pittsburgh, PA 15213

Drexel University
   Philadelphia, PA 19104

Penn State University
   University Park, PA 16802

Wilkes College
   Wilkes-Barre, PA 18766

Tennessee

Middle Tennessee State University
   Murfreesboro, TN 37132

Vanderbilt University
    Nashville, TN 37212

Texas

Rice University
    Houston, TX 77251

Washington

Washington State University
    Pullman, WA 99164

Wisconsin

University of Wisconsin
    Madison, WI 53706

TWO-YEAR COLLEGES OFFERING PROGRAMS RELATED TO WASTE MANAGEMENT

Many two-year colleges offer degree, diploma or certificate programs in water technology, environmental engineering or related areas such as civil engineering technology.

These schools, which may be called junior colleges, community colleges or technical colleges, usually serve a local population. Relatively few have dormitories, meaning the most convenient option is to select a school within driving distance of your home. If this is not possible, consider attending a two-year college in another city or state if you are willing to commute over a long distance or find an apartment or other housing on your own.

To learn what types of programs (if any) related to waste management are offered by any college, consult the school's catalog or contact its office of admissions.

Following is a list of some two-year colleges which offer programs in areas related to waste management. For more details, contact any school in which you are interested and request more information.

Alabama

Gadsden State Community College
    Gadsden, AL 35999

Jefferson State Community College
    Birmingham, AL 35215

Northeast Alabama State Community College
    Rainsville, AL 35986

George C. Wallace State Community College
    Dothan, AL 36303

Arizona

Central Arizona College
    Coolidge, AZ 85228

Cochise College
    Douglas, AZ 85607

Pima Community College
    Tucson, AZ 85709

Phoenix College
    Phoenix, AZ 85013

Rio Salado Community College
    Phoenix, AZ 85003

California

Antelope Valley College
    Lancaster, CA 93536

City College of San Francisco
    San Francisco, CA 94112

Cerritos College
  Norwalk, CA 90650

Citrus Community College
  Glendora, CA 91740

Compton Community College
  Compton, CA 90221

Consumnes River College
  Sacramento, CA 95823

Fullerton College
  Fullerton, CA 92634

Los Angeles Trade and Technical College
  Los Angeles, CA 90015

Modesto Junior College
  Modesto, CA 95350

Palomar College
  San Marcos, CA 92069

Pasadena City College
  Pasadena, CA 91106

Rancho Santiago Community College
  Santa Ana, CA 92706

San Diego City College
  San Diego, CA 92101

San Joaquin Delta College
  Stockton, CA 95207

Ventura College
  Ventura, CA 93003

Santa Rosa Junior College
  Santa Rosa, CA 95401

Colorado

Community College of Denver
  Denver, CO 80218

Red Rocks Community College
  Lakewood, CO 80401

Trinidad State Junior College
  Trinidad, CO 81082

Connecticut

Norwalk State Technical College
  Norwalk, CT 06854

Florida

Brevard Community College
  Cocoa, FL 32922

Broward Community College
  Fort Lauderdale, FL 33301

Indian River Community College
  Fort Pierce, FL 33450

Gulf Coast Community College
  Panama City, FL 32401

Manatee Community College
  Bradenton, FL 32406

Miami-Dade Community College
  Miami, FL 33132

St. Petersburg Junior College
  St. Petersburg, FL 33733

Idaho

North Idaho College
  Coeur D'Alene, ID 83814

Ricks College
  Rexburg, ID 83440

Illinois

College of Du Page
  Glen Ellyn, IL 60137

College of Lake County
  Grayslake, IL 60030

Southeastern Illinois College
  Harrisburg, IL 62946

Waubonsee Community College
  Sugar Grove, IL 60554

Indiana

Indiana Vocational Technical College–Central Indiana
  Indianapolis, IN 46202

Indiana Vocational Technical College–Northeast
  Fort Wayne, IN 46805

Vincennes University
  Vincennes, IN 47591

Iowa

Iowa Western Community College
  Council Bluffs, IA 51501

Kirkwood Community College
  Cedar Rapids, IA 52406

North Iowa Community College
  Mason City, IA 51401

Kansas

Butler County Community College
  El Dorado, KS 67042

Dodge City Community College
  Dodge City, KS 67801

Johnson County Community College
  Overland Park, KS 66210

Kentucky

Jefferson Community College
  Louisville, KY 40402

Maryland

Catonsville Community College
  Baltimore, MD 21228

Cecil Community College
  North East, MD 21901

Massachusetts

Massasoit Community College
  Brockton, MA 02402

Springfield Technical Community College
  Springfield, MA 01105

Michigan

Bay de Noc Community College
  Escanaba, MI 49829

Grand Rapids Community College
  Grand Rapids, MI 49503

Wayne County Community College
  Detroit, MI 48226

Minnesota

Alexandria Technical College
Alexandria, MN 56308

Anoka-Ramsey Community College
Coon Rapids, MN 55433

Normandale Community College
Bloomington, MN 55431

Vermilion Community College
Ely, MN 55731

Mississippi

East Central Junior College
Decatur, MS 39327

Hinds Community College
Raymond, MS 39514

Itawamba Community College
Fulton, MS 38843

Mississippi Delta Junior College
Moorhead, MS 38761

Missouri

Jefferson College
Hillsboro, MO 63050

Nebraska

Metropolitan Community College
Omaha, NE 68103

Southeast Community College
Milford, NE 68405

Nevada

Truckee Meadows Community College
  Reno, NV 89512

New Hampshire

New Hampshire Vocational-Technical College
  Berlin, NH 03570

New Jersey

Burlington County College
  Pemberton, NJ 08068

Glouster County College
  Sewell, NJ 08080

Mercer County Community College
  Trenton, NJ 08690

Union County College
  Crawford, NJ 07882

New Mexico

Eastern New Mexico University–Clovis
  Clovis, NM 88101

New Mexico State University–Carlsbad
  Carlsbad, NM 88220

New York

Onondaga Community College
  Syracuse, NY 13215

Suffolk County Community College
  Selden, NY 11784

North Carolina

Ashville Buncombe Technical College
Ashville, NC 28801

Central Piedmont Community College
Charlotte, NC 28235

Forsyth Technical Community College
Winston-Salem, NC 27103

Sandhills Community College
Pinehurst, NC 28374

Wake Technical College
Raleigh, NC 27603

Ohio

Columbus State Community College
Columbus, OH 43216

Cuyahoga Community College
Cleveland, OH 44115

Edison State Community College
Piqua, OH 45356

Lakeland Community College
Mentor, OH 44060

Stark Technical College
Canton, OH 44720

Oregon

Central Oregon Community College
Bend, OR 97701

Clackamas Community College
Oregon City, OR 97045

Chemeketa Community College
Salem, OR 97309

Linn-Benton Community College
  Albany, OR 97321

Mount Hood Community College
  Gresham, OR 97030

Portland Community College
  Portland, OR 97219

Pennsylvania

Bucks County Community College
  Newton, PA 18940

Community College of Allegheny County–Boyce
  Monroeville, PA 15146

Community College of Allegheny County–South
  West Mifflin, PA 15122

South Carolina

Greenville Technical College
  Greenville, SC 29606

Florence Darlington Technical College
  Florence, SC 29501

Horry-Georgetown Technical College
  Conway, SC 29526

Midlands Technical College
  Columbia, SC 29202

Spartanburg Technical College
  Spartanburg, SC 29303

Sumter Area Technical College
  Sumter, SC 29150

Trident Technical College
  Charleston, SC 29411

York Technical College
    Rock Hill, SC 29730

Tennessee

Chattanooga State Technical Community College
    Chattanooga, TN 37406

State Technical Institute at Memphis
    Memphis TN 38134

Roane State Community College
    Harriman, TN 37748

Texas

Angelina College
    Lufkin, TX 75902

Central Texas College
    Killeen, TX 76541

Houston Community College System
    Houston, TX 77270

Kilgore College
    Kilgore, TX 75662

Midland College
    Midland, TX 79701

Tarrant County Junior College
    Fort Worth, TX 76102

Tyler Junior College
    Tyler, TX 75711

Utah

Salt Lake Community College
    Salt Lake City, UT 84130

Virginia

Central Virginia Community College
  Lynchburg, VA 24502

Lord Fairfax Community College
  Middletown, VA 22645

Mountain Empire Community College
  Big Stone Gap, VA 24219

Southwest Virginia Community College
  Richlands, VA 24641

Tidewater Community College
  Portsmouth, VA 23703

Virginia Western Community College
  Roanoke, VA 24015

Wytheville Community College
  Wytheville, VA 24382

Washington

Centralia College
  Centralia, WA 98531

Grays Harbor College
  Aberdeen, WA 98520

Green River Community College
  Auburn, WA 98002

Everett Community College
  Mount Vernon, WA 98273

Spokane Community College
  Spokane, WA 99207

Walla Walla Community College
  Walla Walla, WA 99362

Wisconsin

Gateway Technical College
  Kenosha, WI 53141

Milwaukee Area Technical College
  Milwaukee, WI 53203

Moraine Park Technical College
  Fond du Lac, WI 54936

Western Wisconsin Technical College
  La Crosse, WI 54602

SELECTED MAGAZINES AND NEWSLETTERS ON WASTE MANAGEMENT

BioCycle
  Box 351
  Emmaus, PA 18049

Fibre Market News
  156 Fifth Avenue
  New York, NY 10010

Recycling Today
  156 Fifth Avenue
  New York, NY 10010

Recycling Times
  Suite 1000
  1730 Rhode Island Avenue
  Washington, D.C. 20004

Resource Recovery Report
  5313 38th Street, N.W.
  Washington, D.C. 20015

Resource Recycling
  P.O. Box 10540
  Portland, OR 97210

Resource Recovery
  National League of Cities
  1301 Pennsylvania Avenue, N.
  Washington, D.C. 20004

Scrap Age
  3615–111 Woodhead Drive
  Northbrook, IL 60062

Scrap Tire News
  Recycling Research, Inc.
  133 Mountain Road
  Suffield, CT 06078

Solid Waste and Power
  HCI Publications
  410 Archibald Street
  Kansas City, MO 64111

Solid Waste Report
  Box 1067, Blair Station
  Silver Spring, MD 20910

Waste Age
  1730 Rhode Island, N.W.
  Suite 512
  Washington, D.C. 20036

Waste Alternatives
  Suite 1000
  1730 Rhode Island Avenue, N.W.
  Washington, D.C. 20004

Waterworld News
  American Water Works Association
  666 West Quincy Avenue
  Denver, CO 80235

World Wastes
  6255 Barfield Road
  Atlanta, GA 30382

VGM CAREER BOOKS

OPPORTUNITIES IN
*Available in both paperback and
hardbound editions*
Accounting
Acting
Advertising
Aerospace
Agriculture
Airline
Animal and Pet Care
Architecture
Automotive Service
Banking
Beauty Culture
Biological Sciences
Biotechnology
Book Publishing
Broadcasting
Building Construction Trades
Business Communication
Business Management
Cable Television
Carpentry
Chemical Engineering
Chemistry
Child Care
Chiropractic Health Care
Civil Engineering
Cleaning Service
Commercial Art and Graphic Design
Computer Aided Design and
    Computer Aided Mfg.
Computer Maintenance
Computer Science
Counseling & Development
Crafts
Culinary
Customer Service
Dance
Data Processing
Dental Care
Direct Marketing
Drafting
Electrical Trades
Electronic and Electrical Engineering
Electronics
Energy
Engineering
Engineering Technology
Environmental
Eye Care
Fashion
Fast Food
Federal Government
Film
Financial
Fire Protection Services
Fitness
Food Services
Foreign Language
Forestry
Gerontology
Government Service
Graphic Communications
Health and Medical
High Tech
Home Economics
Hospital Administration
Hotel & Motel Management
Human Resources Management
    Careers
Information Systems
Insurance
Interior Design
International Business
Journalism
Laser Technology
Law

Law Enforcement and Criminal Justice
Library and Information Science
Machine Trades
Magazine Publishing
Management
Marine & Maritime
Marketing
Materials Science
Mechanical Engineering
Medical Technology
Metalworking
Microelectronics
Military
Modeling
Music
Newspaper Publishing
Nursing
Nutrition
Occupational Therapy
Office Occupations
Opticianry
Optometry
Packaging Science
Paralegal Careers
Paramedical Careers
Part-time & Summer Jobs
Performing Arts
Petroleum
Pharmacy
Photography
Physical Therapy
Physician
Plastics
Plumbing & Pipe Fitting
Podiatric Medicine
Postal Service
Printing
Property Management
Psychiatry
Psychology
Public Health
Public Relations
Purchasing
Real Estate
Recreation and Leisure
Refrigeration and Air Conditioning
Religious Service
Restaurant
Retailing
Robotics
Sales
Sales & Marketing
Secretarial
Securities
Social Science
Social Work
Speech-Language Pathology
Sports & Athletics
Sports Medicine
State and Local Government
Teaching
Technical Communications
Telecommunications
Television and Video
Theatrical Design & Production
Transportation
Travel
Trucking
Veterinary Medicine
Visual Arts
Vocational and Technical
Warehousing
Waste Management
Welding
Word Processing
Writing
Your Own Service Business

CAREERS IN Accounting; Advertising;
Business; Communications; Computers;
Education; Engineering; Health Care;
High Tech; Law; Marketing; Medicine;
Science

CAREER DIRECTORIES
Careers Encyclopedia
Dictionary of Occupational Titles
Occupational Outlook Handbook

CAREER PLANNING
Admissions Guide to Selective
    Business Schools
Career Planning and Development for
    College Students and Recent
    Graduates
Careers Checklists
Careers for Animal Lovers
Careers for Bookworms
Careers for Culture Lovers
Careers for Foreign Language
    Aficionados
Careers for Good Samaritans
Careers for Gourmets
Careers for Nature Lovers
Careers for Numbers Crunchers
Careers for Sports Nuts
Careers for Travel Buffs
Guide to Basic Resume Writing
Handbook of Business and
    Management Careers
Handbook of Health Care Careers
Handbook of Scientific and
    Technical Careers
How to Change Your Career
How to Choose the Right Career
How to Get and Keep
    Your First Job
How to Get into the Right Law School
How to Get People to Do Things
    Your Way
How to Have a Winning Job Interview
How to Land a Better Job
How to Make the Right Career Moves
How to Market Your College Degree
How to Prepare a *Curriculum Vitae*
How to Prepare for College
How to Run Your Own Home Business
How to Succeed in Collge
How to Succeed in High School
How to Write a Winning Resume
Joyce Lain Kennedy's Career Book
Planning Your Career of Tomorrow
Planning Your College Education
Planning Your Military Career
Planning Your Young Child's
    Education
Resumes for Advertising Careers
Resumes for College Students & Recent
    Graduates
Resumes for Communications Careers
Resumes for Education Careers
Resumes for High School Graduates
Resumes for High Tech Careers
Resumes for Sales and Marketing Careers
Successful Interviewing for College
    Seniors

SURVIVAL GUIDES
Dropping Out or Hanging In
High School Survival Guide
College Survival Guide

VGM Career Horizons
a division of *NTC Publishing Group*
4255 West Touhy Avenue
Lincolnwood, Illinois 60646-1975